Green Engineering and Sustainability in the Pharmaceutical Industry

Topical Conference at the 2008 AIChE Annual Meeting

Philadelphia, Pennsylvania
16 - 21 November 2008

ISBN: 978-1-61567-202-8

Printed from e-media with permission by:

Curran Associates, Inc.
57 Morehouse Lane
Red Hook, NY 12571

Some format issues inherent in the e-media version may also appear in this print version.

Printed by Curran Associates, Inc. (2009)

For permission requests, please contact AIChE
at the address below.

AIChE
3 Park Avenue
New York, NY 10016-5991

Phone: (203) 702-7660
Fax: (203) 775-5177

www.aiche.org

Additional copies of this publication are available from:

Curran Associates, Inc.
57 Morehouse Lane
Red Hook, NY 12571 USA
Phone: 845-758-0400
Fax: 845-758-2634
Email: curran@proceedings.com
Web: www.proceedings.com

TABLE OF CONTENTS

13a Business Case: Pharmaceutical Sustainability

Ann Lee-Jeffs, Johnson & Johnson, 410 George St, New Brunswick, NJ 08901 and John Leazer, Merck Process Research, Merck, Rahway, NJ 07065.

Text Not Available.

13b Model Based Solvent Selection with NRTL-SAC Including Use of Chromatographic Retention Times as Model Input

Alexander J. Marchut, Bristol Myers Squibb, New Brunswick, NJ 08903, Olav Lyngberg, Process Research and Development, Bristol-Myers Squibb, PO Box 191, New Brunswick, NJ 08903, and Chau-Chyun Chen, Aspen Technology, Inc., 10 Canal Park, Cambridge, MA 02141.

Model based solvent selection can help to point out green alternatives to environmentally unfriendly solvents by examining a much wider range of solvent space than is possible with even high throughput measurements. Use of a model for solvent selection can also help to cut down the number of experiments required for solvent selection so that less waste is generated. The NRTL SAC (Non-Random Two Liquid Segment Activity Coefficient) model is a parameter based method for estimating solubilities in different solvents which can be used for solvent selection. It has been widely applied at Bristol Myers Squibb in over 100 compounds for this purpose. Typically, five to ten solubility measurements are used as input to the model and 0.25 to 0.5 grams of material is required for these measurements. In order to provide solubility predictions earlier in development using less material and further cut down on waste generation, a model for predicting solubility data using chromatographic retention times as model input has been developed based on the NRTL-SAC solubility model. Using the model, NRTL-SAC model parameters were established for twelve chromatographic stationary phases from retention times of molecules with known NRTL-SAC model parameters. Based on these model parameters for the stationary phases, retention times were used to find the previously unknown model parameters for a set of test molecules which were then used to predict their solubilities. These predictions as well as case studies in model-based solvent selection will be presented.

13c Piperylene Sulfone: Fully Recyclable Diprotic Apolar Solvent for Benign Pharmaceutical Processes

Charles A. Eckert[1], Charles L Llotta[1], and Douglas P. Kjell[2]. (1) School of Chemical & Biomolecular Engineering and Specialty Separations Center, Georgia Institute of Technology, 311 Ferst Drive, NW, Atlanta, GA 30332-0100, (2) Chemical Process Research & Development, Lilly, Lilly Research Laboratories, Lilly Corporate Center, Indianapolis, IN 46285

Many pharmaceutical intermediates require polar aprotic solvents for dissolution, but solvent removal and product purification can yield undesirable environmental effects. Typical reactions in polar aprotics include nucleophilic displacements, amide formation, and metal mediated couplings, and it is not unusual to need large amounts of water to isolate product from the polar aprotic – either by precipitation or extraction. We have excerpted 19 patents from Organic Process Research & Development, 2008(1) and found that six of these explicitly mention a polar aprotic solvent: Four are nucleophilic displacements (one each ether formation, fluorination, alklyation, thioether formation), and two are amide formations. Of these six, two cite as solvent DMSO and four DMF.

In fact, dipolar, aprotic solvents such as DMSO and DMF are excellent for organic reactions with inorganic salts because they are inexpensive and dissolve disparate materials. But their high boiling points preclude benign separation and recycle. We report piperylene sulfone, a labile DMSO substitute, which undergoes a retro-cheletropic reaction into volatile species under modest conditions, and which can easily be collected and converted back into piperylene sulfone via a concerted reaction for re-use. This provides the opportunity for facilitating recovery and recycling of both catalyst as well as the solvent itself.

We have already investigated the use of piperylene sulfone to couple separation and solvent and catalyst recycle with many reactions. In nucleophilic substitution reactions it gives rates comparable to those in DMSO. We show TEMPO oxidations of alcohols in piperylene sulfone with higher turn over frequency than DMSO and multiple recyles. Other examples include asymmetric transfer hydrogenation in piperylene sulfone and sulfone induced in situ acid catalysis. Here we show new opportunities for this solvent in the pharmaceutical industry.

Self-condensation of acetophenone to dypnone: synergism of microwave and solid acid catalysts

Ganapati D. Yadav* and Ginish George

Department of Chemical Engineering,

University Institute of Chemical Technology,

University of Mumbai,

Matunga, Mumbai - 400 019, INDIA

Email: gdyadav@yahoo.com, gdyadav@udct.org

Solid acid catalysts are widely used for the development of green technology to replace the traditional polluting catalysts in organic process industries employing alkylation, acylation, isomerization, nitration, condensation, cracking, and esterification, etc. [1,2] The activity and selectivities of different solid acid catalysts such as zeolites,[3,4] modified clays and heteropoly acids,[5-12] sulphated zirconia,[13-16] ion exchange resins[17-19] and UDCaT series of catalysts[20-26] have been successfully evaluated in our laboratory for developing green acylation, alkylation, condensation, cyclisation, esterification, etherification, isomerization, oligomerization, and nitration. The acronym UDCaT stands for the series of catalysts developed in our laboratory (named after University Institute of Chemical Technology).

Dypnone is a useful intermediate for the production of a large range of compounds. It has been used as a sunscreen lotion, softening agent, plasticizer and perfumery base. Dypnone has been prepared conventionally by the action of sodium ethoxide, aluminum bromide, phosphorous pentachloride, aluminum triphenyl, zinc diethyl, calcium hydroxide, anhydrous hydrogen chloride, anhydrous hydrogen bromide, aluminum chloride, aluminum *tert*-butoxide or hydrogen fluoride on acetophenone as described in the earlier literature. It has also been obtained by the action of aniline hydrochloride on acetophenone, followed by treatment with hydrochloric acid. In addition, $AlCl_3$ being a strong Lewis acid also catalyzes other undesirable side reactions such as polycondensation of acetophenone to a mixture of other products.[27-30] Yatluk *et al.*[31] studied the condensation of acetophenone by using titanium ethoxide. Heterogeneous catalysts like zeolites and sulphated metal oxides were also employed in this system, but, however were more energy intensive and took longer times for equivalent conversions.[32-34]

High-speed synthesis with microwaves has attracted a considerable amount of attention in recent years.[35,36] Not only is direct microwave heating able to reduce chemical reaction times from hours to minutes, but it is also known to reduce side reactions, increase yields, and improve reproducibility.

The current work gives an insight into the self-condensation of acetophenone under microwave irradiation by using novel nanocatalyst Cs-DTP/K-10 to get dypnone (1,3-diphenyl-2-buten-1-one) as the major product including choice of catalyst and kinetic modeling.

The reaction was carried out in a 120 mL capacity glass reactor of 4 cm i.d. equipped with four equally spaced baffles and six bladed turbine impeller. The reaction conditions were controlled by using a microwave reactor setup (Discover, CEM-SP 1245 model, CEM Corp. USA), which was maintained at a desired temperature with an accuracy of ±0.5 K, in which the reaction assembly was immersed. Standard experiments were carried out by taking 0.16 mol of acetophenone, at a temperature of 413 K, and 0.10 g/cm^3 catalyst loading. 1.07 g of n-decane was used as an internal standard. All catalysts were dried at 393 K for 3 h before use. The reaction mixture was allowed to reach the desired temperature and the initial sample collected. Agitation was then commenced at a known speed. Samples were withdrawn periodically and analyzed by gas chromatography (GC). The conversion is defined as the percentage of acetophenone transformed. The products were confirmed by GC-MS.

The reaction scheme is as shown below:

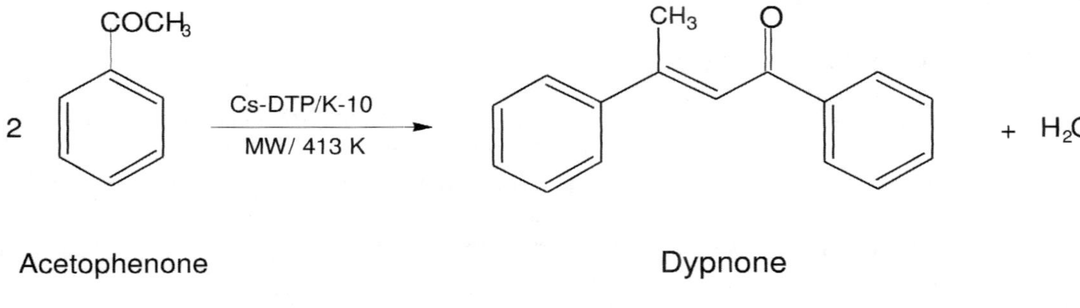

Analysis of the reaction mixture was performed by GC (Chemito, model 8510) by using flame ionization detector and a stainless steel column (3.25 mm diameter and 4 m length) packed with a stationary phase of 10% SE-30 supported on chromosorb-WHP. The conversion was based on the disappearance of acetophenone peak with respect to the internal standard. Selectivity towards the products was analyzed by using GC (Chemito Model 1000, FID detector) with a BP1 capillary column (0.22 mm x 25 m).

Results and Discussion

Various solid acid catalysts were used to assess their efficacy in this reaction. A 0.10 g/cm^3 loading of catalyst based on the volume of the reaction mixture was employed at 413 K and a speed of agitation of 1000 rpm. The catalysts were UDCaT-5, Cs-DTP/K-10, 20% w/w $Cs_{2.5}H_{0.5}PW_{12}O_{40}$/HMS, and UDCaT-6. The order of activity was as follows:

Cs-DTP/K-10 (most active)> UDCaT-5> Cs-DTP/HMS > UDCaT-6 (least)

Novel nanocatalyst cesium substituted dodecatungstophosphoric acid supported on K-10 clay (Cs-DTP/K-10) showed highest activity comparing to other catalysts (Fig.1). Nadir *et al.*[36] reported the synergistic combination of microwave irradiation and K-10 clay for organic synthesis. The narrow pore sizes of K-10 clay based catalysts show good selectivity towards desired product. Further experiments were conducted with Cs-DTP/K-10 due to its excellent stability and reusability under microwave condition. A brief characterization of this catalyst is presented.

Self-condensation of acetophenone was performed in the presence of microwave irradiation as well as conventional heating under otherwise similar conditions (Fig. 2). It was found that the overall conversion as well as rate of reaction was much higher under microwave vis-à-vis the conventional heating. This shows that the effect may not be purely thermal. Further, only microwave irradiation was carried out without any catalyst loading, there was no reaction. Thus, there is a definite synergism between solid acid catalyst and microwave irradiation.

The effect of speed of agitation was studied in the range of 800-1200 rpm at a catalyst loading 0.10 g/cm^3 at 413 K . There was no significant change in the rate and conversion patterns, which was indicative of the absence of any resistance to external mass transfer of acetophenone to the external surface of the catalysts. However, all further reactions were carried out at a speed of 1000 rpm. Theoretical calculations were also done to establish that there was absence of external mass transfer resistance.

The effect of catalyst loading was studied over range of 0.03-0.10 g/cm^3. In the absence of mass transfer resistance, the rate of reaction was directly proportional to catalyst loading based on the entire liquid phase volume. This indicates that, as the catalyst loading increased the conversion of acetophenone increases, which is due to proportional increase in the number of active sites. Further reactions were carried out with 0.10 g/cm^3 catalyst loading used in the standard reaction.

The effect of temperature on conversion was studied under otherwise similar conditions at 383, 393, 403 and 413 K, respectively. It was observed that the conversion increased with temperature. This would suggest a kinetically controlled mechanism. The initial rates of reaction were calculated at different temperatures and the Arrhenius plots had made to determine the energy of activation. It was found to be 16.35 kcal/mol, which is an indication of the overall rate being controlled by intrinsic kinetics.

The catalyst reusability was studied four times, including the use of fresh catalyst. After each run, the catalyst was filtered, washed with methanol, dried at 393 K for 3h, and weighed before using in the next batch of reaction. There was some attrition of catalyst particles during agitation. In a typical batch reaction, there was an inevitable loss of particles during filtration due to attrition. Although the catalyst was washed with methanol after filtration to remove all adsorbed reactant and products, there was still a possibility of retention of some amount of adsorbed reactant and products which might cause the blockage of active sites of the catalyst. The actual amount of catalyst used in the next batch, was almost 5% less than the previous batch. The volume of the reaction mixture was adjusted to make the catalyst loading 0.10 g/cm^3, the standard value. The catalyst was stable under microwave irradiation and there was only a marginal decrease in conversion.

The above results were used to build a kinetic model. A model based on two catalytic sites was proposed When the adsorption constants of all species are very weak, it is a typical second order reaction (Fig 3).

Conclusion

The current work has addressed the self-condensation of acetophenone to dypnone at 413 K over variety of solid acid catalysts such as Cs-DTP/K-10, UDCaT-5, UDCaT-6 and $Cs_{2.5}H_{0.5}PW_{12}O_{40}$/HMS. Cs-DTP/K-10 was found to be the most active, stable and reusable catalyst under microwave conditions, gives very high selectivity towards *trans*-dypnone (92%). The effects of various parameters on the rates over Cs-DTP/K-10 were discussed. A positive effect of microwave irradiation on the rate of reaction was observed in comparison with conventional heating. A second order rate equation fits the data well. The apparent activation energy was estimated as 16.35 kcal/mol.

References

[1] R.A. Sheldon, *Chemistry & Industry*, 1997, **1**, 12.

[2] K. Wilson and J.H. Clark, *Pure Appl. Chem.*, 2000, **72**(7), 1313-1319

[3] G.D. Yadav and A.A. Pujari, *Can. J. Chem. Eng.*, 1999, **77**(3), 489-496

[4] G.D. Yadav and D.V. Satoskar, *J. Chem. Tech. Biotech.*, 1997,**69**(4), 438-450

[5] G.D. Yadav, *Cat. Surveys Asia*, 2005, **9**(2), 117-137

[6] G.D. Yadav and N. Kirthivasan, *J. Chem. Soc. Chem. Commun.*, 1995, 203-204

[7] G.D. Yadav and V. V. Bokade, *Appl. Catal. A: Gen.*, 1996, **147**, 299-315

[8] G.D. Yadav and N.S. Asthana, *Ind. Eng. Chem. Res.*, 2002, **41**, 5565-5575

[9] G.D. Yadav, N.S. Asthana and V S. Kamble, *App. Catal. A: Gen.*, 2003, **240**, 53-69

[10] G.D. Yadav and N.S. Asthana, *App.Catal. A: Gen.*, 2003, **244**, 341-357

[11] G.D. Yadav, N.S. Asthana and V. S.Kamble, *J. Catal.*, 2003, **217**(1), 88-99

[12] G.D. Yadav and H.G. Manyar, *Micropor. Mesopor. Mater.*, 2003, **63**(1-2), 85-96

[13] P.S. Kumbhar and G.D. Yadav, *Chem. Eng. Sci.*, 1989, **44**, 2535-2544

[14] G.D. Yadav and J.J. Nair, *Micropor. Mesopor. Mater.*, 1999, **33**, 1-48

[15] G.D. Yadav and P. Ramesh, *Can. J. Chem. Engg.*, 2000, **78**, 323-327

[16] G.D. Yadav and S. Sengupta, *Org. Proc. Res. Dev.,* 2002, **6**, 256-262

[17] G.D. Yadav and S.P. Nalawade, *Chem. Eng. Sci.,* 2003, **58**, 2573-2585

[18] G.D. Yadav and M.S.M.M. Rahuman, *Org. Proc. Res. Dev.*, 2002, **6**, 706-713

[19] G.D. Yadav and P.K. Goel, *Green Chem.,* 2000, **2**(2), 71-77

[20] G.D. Yadav, M.S. Krishnan, A.A. Pujari, N.S. Doshi and M.S.M.M. Rahuman, *US Pat.,* 6 204 424 B1, 2001

[21] G.D. Yadav, J.J. Nair and V. Narendra, *US Pat.*, 6 177 596 B1, 2001

[22] G.D. Yadav and J.J. Nair, *Chem. Comm.,* 1998, 2369-2370

[23] G.D. Yadav and A.D. Murkute, *Adv. Synth. Catal.*, 2004, **346**, 389-394

[24] G.D. Yadav and A.D. Murkute, *Langmuir,* 2004, **20**, 11607-11619

[25] G.D. Yadav and A.D. Murkute, *J. Catal.*, 2004, **224**, 218-223

[26] G.D. Yadav and A.D. Murkute, *J.Phy.Chem. A,* 2004, **108** (44), 9557-9566

[27] H. Adkins and F.W. Cox, *J. Am. Chem. Soc.,* 1938, **60**, 1151-1159

[28] W. Wayne and H. Adkins, *Org. Synth. Coll. III,* 1955, 367

[29] N.O. Calloway and L.D. Green, *J. Am. Chem. Soc.*, 1937, **59**, 809-811

[30] J.H. Simons and E.O. Ramler, J. *Am. Chem. Soc.*, 1943, **65**, 1390-1392

[31] Y.G. Yatluk, V.Y. Sosnovskikh and A.L. Suvorov, *Russian J. Org. Chem.*, 2004, **40**(6), 763-765

[32] P.B. Venuto, *Micropor. Mater.*, 1994, **2**, 297-411

[33] C. Venkatesan and A.P. Singh, *J. Mol. Catal. A: Chem.*, 2002, **181**, 179-187

[34] M. Selvaraj, P.K. Sinha and A. Pandurangan, *Micropr. Mesopor. Mater.*, 2004, **70**, 81-91

[35] C.O. Kappe, *Angew. Chem. Int. Ed.*, 2004, **43**, 6250-6284

[36] U.K. Nadir and A. Singh, *Tet. Lett.*, 2005, **46**, 2083-2086

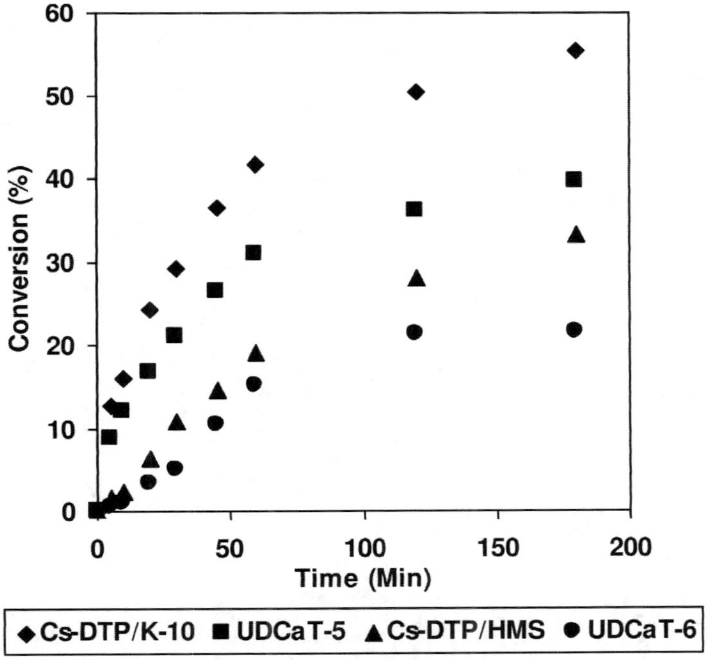

Fig. 1. Effect of various catalysts: acetophenone 0.16 mol, catalyst loading 0.10 g/cm^3, temperature 413 K, speed of agitation 1000 rpm

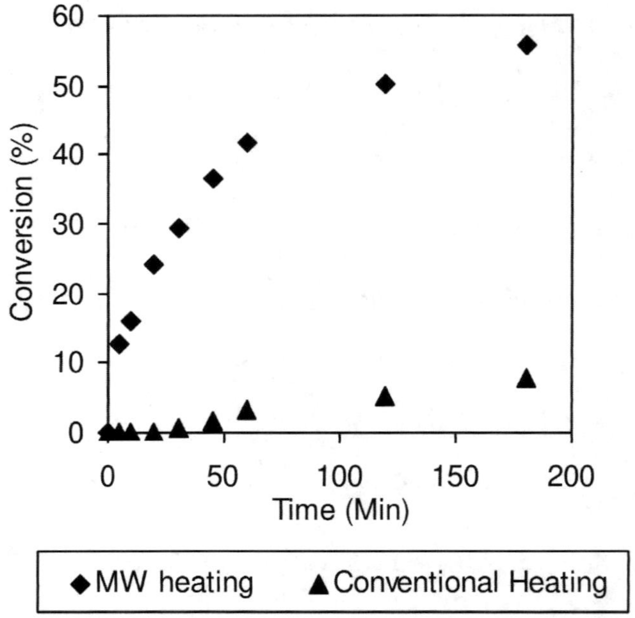

Fig. 2. Microwave vs. conventional heating: acetophenone 0.16 mol, catalyst loading 0.10 g/cm^3, temperature 413 K, speed of agitation 1000 rpm

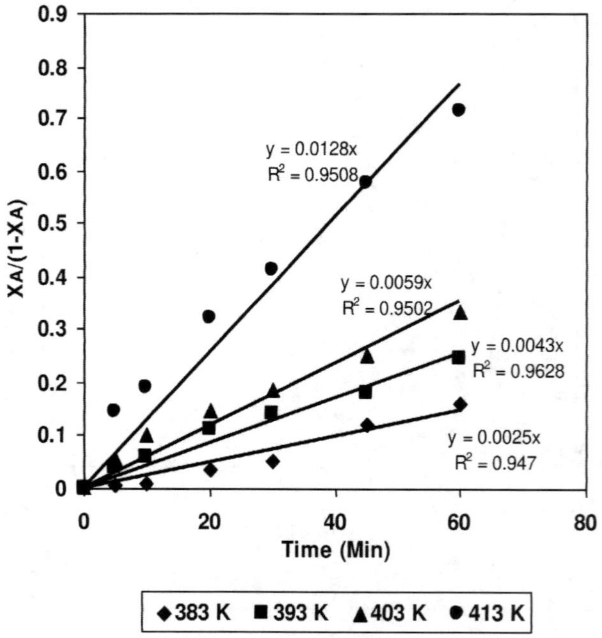

Fig. 3. Validation of the mathematical model at different temperature

13e Homogeneous Catalysis Coupled with Benign Separation In Polyethylene Glycol-Organic-Carbon Dioxide Tunable Solvent Systems

Vittoria M. Blasucci[1], Megan Donaldson, Charles A. Eckert, and Charles L. Liotta[2]. (1) School of Chemical & Biomolecular engineering, Georgia Institute of Technology, 311 Ferst Dr., Atlanta, GA 30332, (2) Chemistry and Biochemistry, Georgia Institute of Technology, PO Box 330834 Georgia Tech Station, Atlanta, GA 30332

In this work, we used a polyethylene glycol (PEG)-dioxane-carbon dioxide solvent system for sustainable coupling of homogeneous catalytic reactions with facile heterogeneous separations. At low molecular weights (i.e. 200-600), PEG is non-toxic and has negligible vapor pressure, making it an attractive alternative to traditional volatile organic solvents. By applying carbon dioxide pressure to the system, we can split the PEG-dioxane mixture into two phases, allowing for catalyst recycle in the PEG phase and product recovery in the gas expanded dioxane phase. These carbon dioxide tunable solvent systems enable improved reaction rates, catalyst recycle, and facile product recovery. Specifically, we have used these systems for palladium catalyzed carbon-oxygen coupling of aryl halides to produce both phenols and asymmetric aromatic ethers. These molecules are prevalent in both pharmaceutical and chemical industries, but often production requires harsh and hazardous reaction conditions. In the phenol case, carbon dioxide plays a dual role by acting as the neutralization media via in-situ generation of carbonic acid. Carbonic acid is formed from the equilibrium between water and carbon dioxide and is reversible upon venting. This benign work-up step eliminates conventional acidifications which create large quantities of waste and inactivate the catalyst for further use. Initial results show high selectivity to phenol or ether, enhanced reaction rates, and favorable product partitioning into the carbon dioxide expanded dioxane phase.

13f Collaboration to Deliver a Solvent Selection Guide for the Pharmaceutical Industry

Caireen R. Hargreaves, Safety, Health and Environment Global Technical Support, AstraZeneca, Alderley Park, Macclesfield, United Kingdom

During pharmaceutical process development, solvent selection is a key factor in determining the sustainability of future commercial production methods due to the significant amount of solvents used. Previous benchmarking studies within the industry have shown that solvents account for around 50% of the total materials used in the manufacture of active pharmaceutical ingredients.

In recent years a number of pharmaceutical companies have independently developed tools enabling their scientists to consider relevant green criteria when making solvent selection decisions. These tools range from simple guides for early research chemistry departments to more complex resources for chemists and chemical engineers working on process development towards commercial manufacturing routes. In 2008 the American Chemical Society Green Chemistry Institute® Pharmaceutical Roundtable (ACS GCIPR) adopted an objective to collaborate and use the experiences of the individual member companies to develop a global solvent selection guide for use across the industry and beyond.

In the first phase of this work a set of environmental, human health and safety criteria together with a scoring methodology were defined. A number of key solvents used by the industry were identified and scored against the criteria to form the basis of the guide allowing easy comparison of solvent profiles.

The work contributes to the mission of the ACS GCIPR to catalyse the implementation of green chemistry and engineering in the pharmaceutical industry globally by delivering a practical tool that can be used by scientists to make informed solvent selection decisions with due consideration to sustainability aspects.

Web Page: www.acs.org/gcipharmaroundtable

13g Solvents In Organic Synthesis: Replacement and Multi-Step Reaction Systems

Rafiqul Gani, CAPEC, Department of Chemical and Biochemical Engineering, Technical University of Denmark, Søltofts Plads, Building 229, 2800 Kgs. Lyngby, Denmark, Concepcion Jimenez-Gonzalez, GSK, CEHS, 5 Moore Dr, Research Triangle Park, NC 27709, David Constable, Cehs, GlaxoSmithKline, 1 Franklin Plaza, Philadelphi, PA 19103, and Muhammad Shafique Bashir, CAPEC, Department of Chemical Engineering,, Technical University of Denmark,, Dk-2800, Lyngby,, Dk-2800, Denmark.

Solvents are widely used as reaction media in the chemical, fine chemical and pharmaceutical industries, but they present numerous environmental, health and safety (EHS) challenges that need to be managed and are subject to increasing regulatory scrutiny. The above issues, together with the principles of green chemistry, highlight the need to minimize and optimize the use of organic solvents as much as possible. One important step in optimizing organic solvent use is the identification of suitable 'greener' solvents that can help to minimize the environmental, health and safety concerns during design and commercial manufacture of chemical products. A method for selecting appropriate 'greener' solvents for the promotion of a class of organic reactions has been previously developed by Gani et al. This method employs estimates of thermodynamic properties to generate a knowledge base of reaction- and solvent-related properties that directly or indirectly influence a given reaction. In this work, the methodology has been extended through its application to several cases of increased process complexity involving multi-step reaction systems and solvent substitution for specific reaction steps in existing processing systems. The problems were formulated using the original methodology guidelines and incorporate knowledge of industrial practice in the currently available computer-aided tools for solvent design and property estimation.

This presentation will describe the results obtained for a single reaction solvent screening problem, several solvent replacement problems and a multi-stage reaction system. In each example a list of solvent candidates is generated so they may be further investigated experimentally.

107a Tools and Metrics for Environmental Sustainability

Raymond L. Smith and Michael A. Gonzalez. US EPA, National Risk Management Research Lab, 26 W. Martin Luther King Dr., Cincinnati, OH 45268

Within the U.S. Environmental Protection Agency's Office of Research and Development the National Risk Management Research Laboratory has been developing tools to help design and evaluate chemical processes with a life cycle perspective. These tools include the Waste Reduction (WAR) Algorithm, which is useful in the chemical process industries for evaluating eight environmental impact categories with over 2000 chemicals in its database. Applied inside WAR is a tool for toxicity prediction which can be employed on a very wide variety of chemicals. Also developed in the Laboratory is the impact assessment tool TRACI (Tool for the Replacement and Assessment of Chemical and other environmental Impacts). TRACI can be used in life cycle assessments or chemical process evaluations to evaluate the environmental fate and transport of 500+ chemicals. Both the WAR and TRACI tools are used in environmental evaluations of chemical processes by the developing tool, GREENSCOPE. GREENSCOPE considers four areas to provide metrics of the sustainability of a process or technology change. These four areas are economics, environment, efficiency and energy use. The many metrics in these four areas provide multiple perspectives on the multifaceted objective of sustainability in one tool. As this tool is further developed and used it will link these relevant metrics to sustainable practices. Examples of organic chemistries will be discussed with respect to these tools and metrics.

107b Land Use Change Effects on Life Cycle Greenhouse Gas Emissions of Bio-Based Fuels and Pharmaceutical Products

David R. Shonnard, Department of Chemical Engineering, Michigan Technological University, 1400 Townsend Dr., Houghton, MI 49931

Abstract

Over the past decade, the production of corn ethanol and of biodiesel from plant oils such as soybean, rapeseed, and palm oils has experienced rapid growth and increased international market demand. This demand for biofuel feedstocks is beginning to affect decision by farmers to produce crops for fuel as opposed to food. This competition for farm land to satisfy simultaneous demands for fuel and for food, as well as other factors such as higher costs for energy, fertilizers, and other farm chemicals, has resulted in higher commodity food prices on a global scale (World Bank 2008). In addition, biofuels produced from agricultural lands that compete directly with food production have been implicated with global-scale land use change "carbon debt" (Searchinger et al. 2008, Fargione et al. 2008). Biofuels directives in the European Union and in the USA are considering requirements that biofuels producers and biomass feedstock suppliers demonstrate that the entire life cycle of biofuels are sustainability.

The purpose of this presentation is to review and explore the land use change issue from the standpoint of both direct and indirect effects. The primary focus will be on biofuels due their potential for large-scale land use change. The presentation will conclude with a discussion of the potential impacts of the pharmaceutical industry in the US on the land use change issue.

References

Fargione, J., Hill, J., Tilman, D., Polasky, S., Hawthorne, P. 2008, Land clearing and the biofuel carbon debt, Science, www.sciencexpress.org / 7 February 2008 / Page 1 / 10.1126/science.1152747

Searchinger, T, Heimlich, R., Houghton, R.A., Dong, F., Elobeid, A., Fabiosa, J., Tokgoz, S., Hayes, D., Yu, T-Y. 2008, Use of U.S. croplands for biofuels increases greenhouse gases through emissions from land use change, Science, www.sciencexpress.org / 7 February 2008 / Page 1 / 10.1126/science.1151861

World Bank, 2008, Rising Food Prices: Policy Options and World Ban Response, April 9, 2008, http://web.worldbank.org/WBSITE/EXTERNAL/NEWS/0,,contentMDK:21722688~pagePK:64257043~piPK:437376~theSitePK:4607,00.html

107c Application of Chemical Engineering Tools for Development of Life Cycle Inventories In the Chemical Processing Industry

Evan M.H. Griffing, Environmental Clarity, 1575 Alameda St, Pomona, CA 91768

Traditional consensus based life cycle inventory (LCI) generation is time consuming and often yields inconsistent data. A review of commercial databases shows life cycle inventories that are not balanced on an atom level. In addition, energy inputs are sometimes inconsistent with thermodynamic laws. These inconsistencies can be caused by allocation of inventories in multiple product processes or by averaging several field measurements of inputs and outputs.

Chemical engineers have the tools to generate internally consistent LCI data from publicly available information using the process-based methodology. Two strengths of this method are generating reports that are easily peer reviewed and doing so relatively quickly. These characteristics are very important to the fine chemical and pharmaceutical industries, where life-cycle supply chains quickly grow large in scale.

We will discuss our experience generating a database commodity chemicals, solvents, catalysts, etc. using the process-based system. Issues specific to LCA include model complexity, allocation, presentation of cradle to gate data, and systems for managing data. A case study will be used to illustrate these concepts.

Web Page: www.environmentalclarity.com

107d Lessons Learned through Measuring Green Chemistry Performance: The Pharmaceutical Experience

John M. Kindervater, Eli Lilly and Company, Lilly Corporate Center, Indianapolis, IN 46285, Richard Henderson, Cehs, GlaxoSmithKline, Park Road, Ware, Hertfordshire, SD12 ODP, United Kingdom, and Julie Manley, ACS Green Chemistry Institute, 1155 Sixteenth St, NW, Washington, DC 20036.

The ACS Green Chemistry Institute® Pharmaceutical Roundtable is a global, collaborative organization currently comprised of nine international pharmaceutical companies whose mission is to catalyze the implementation of green chemistry and engineering in the pharmaceutical industry. Membership is open to all pharmaceutical research, development, and manufacturing companies and the roundtable will be strongest when all global pharmaceutical corporations are members. The strategic aims of the roundtable are to inform and influence the green chemistry research agenda to help meet the challenges of the pharmaceutical industry; to develop common tools for use across the pharmaceutical industry; to provide resources for education about green chemistry focused on the pharmaceutical industry; and to be a forum for cooperation and collaboration between the member companies to embed green chemistry and science into everyday operations and processes.

One of the major green chemistry challenges for the pharmaceutical industry is to decrease the process mass intensity (the amount of material used to make a drug per kg of drug) of the manufacturing processes used to produce Active Pharmaceutical Ingredients (API). To begin addressing this challenge, ACS GCI Pharmaceutical Roundtable members have developed a common process mass intensity metric that allows data from each company to be compared on a transparent and equitable basis. Data from member companies has been collected as part of a benchmarking exercise and has yielded a representative snapshot of the process mass intensities observed for those compounds under development at each company.

This talk will focus on an explanation of process mass intensity and will provide results of the benchmarking exercise between member companies.

Web Page: www.acs.org/gcipharmaroundtable

107e Life Cycle Inventory for Production of Vancomycin HCl and the Effect of Solvent Choices

Celia S. Ponder, Chemical and Biomolecular Engineering, North Carolina State University, 911 Partners Way, Raleigh, NC 27695 and Michael R. Overcash, University of California, Davis.

This paper presents the cradle-to-gate life cycle inventory for the manufacture of the active pharmaceutical ingredient (API) vancomycin hydrochloride, a drug used in the treatment of methicillan-resistant Staphylococcus aureus (MRSA). A general fermentation process using Amycolatopsis orientalis to make vancomycin is used, with purification by adsorption and crystallization. After fermentation, the vancomycin solution is filtered and passed through adsorption columns to remove color and salt impurities. Crystallization can then be performed using various solvents and chemicals such as sodium hydroxide, ethanol, isopropanol, phosphoric acid, potassium phosphate, hydrochloric acid, acetone, ammonium chloride, and ammonium hydroxide to purify and convert the vancomycin to vancomycin hydrochloride. Three different crystallization processes are studied to show which solvents are the greener choice based on a life cycle analysis. The scope of the inventory is from the factory to the supply chain chemicals, and the inventory has been generated using the design-based methodology.

107f Resource Efficiency and Fingerprinting Metrics through Exergetic Life Cycle Analysis

Jo Dewulf[1], Geert Van der Vorst[1], Herman Van Langenhove[1], and Wim Aelterman[2]. (1) Ghent University, Ghent, 9000, Belgium, (2) Johnson & Johnson PRD, Janssen Pharmaceutica nv, Beerse, 2340, Belgium

Environmental Performance of Technology is undoubtedly a critical issue in developing sustainable technology. Emphasis for a long time now has been on reducing the emission pattern of processes. It is not surprising that traditional LCA pops up here as a mature tool for the assessment of improvements: it contains several emission impact categories such as global warming, ozone depletion, toxics emission, acidification etc.

However, at the other end of the production chain technology puts pressure on the environment through the extraction of resources. Nowadays, with an obvious depletion of some of the natural resources through a growing worldwide demand, technology will undoubtedly be forced to deal in an efficient way with the resources and maybe it will also need to shift the pattern of the resources it takes in, for example more renewables.

Metrics with respect to resource intake, resource efficiency and resource fingerprinting are not that mature as emission assessment tools. Nevertheless, it is obvious that thermodynamics through its exergy concept brings strong opportunities (Dewulf et al., 2008). It gives a consistent basis to quantify all kind of resources: organic and inorganic, fuel and feedstock, renewable and non-renewable, energy and material, ... Through combination with a life cycle approach into Exergetic Life Cycle Analysis, it may be one of the best available resource assessment tools.

The presentation here brings two main parts. In a first part it learns the potential of exergy and especially exergetic life cycle assessment as a resource assessment tool. Second, through two specific cases in the pharmaceutical industry (Plant of Johnson & Johnson PRD, Janssen Pharmaceutica, Beerse, Belgium), it is demonstrated what insights it brings through looking at the full supply chain and through looking at materials and energy, feedstocks and utilities. The two cases are a separation through chromatography versus crystallisation, and supercritical fluid chromatography versus high performance liquid chromatography.

References:

1.J. Dewulf, G. Van der Vorst, A. Aelterman, B. De Witte, H. Vanbaelen and H. Van Langenhove. Integral resource management by exergy analysis for the selection of a separation process in the pharmaceutical industry. Green Chemistry, 9 (2007) 785-791.

2. J. Dewulf, H. Van Langenhove, B. Muys, S. Bruers, B.R. Bakshi, G.F. Grubb, D.M. Paulus, E. Sciubba. Exergy: its potential and limitations in Environmental Science and Technology. Environmental Science and Technology 42 (2008) 2221-2232.

107g GlaxoSmithKline's Experiences with Life Cycle Assessment

Concepcion Jimenez-Gonzalez, GSK, CEHS, 5 Moore Dr, Research Triangle Park, NC 27709 and David Constable, Cehs, GlaxoSmithKline, 1 Franklin Plaza, Philadelphi, PA 19103.

GlaxoSmithKline has investigated life cycle impacts in the pharmaceutical supply chain for more thana decade, incorporating life cycle inventory and assessment (LCI/A) tools within metrics, tools and decision systems within the organization to drive towards more sustainable business pactices. This presentation intends to briefly cover GSK's experiencies in developing, applying and embedding LCI/A concepts, metrics and tools into business process.

Title: Continuous Processing in Pharmaceutical Process Development

Authors: Chenchi Wang and Thomas LaPorte

As an alternate technology in pharmaceutical process development, continuous processing is applied to various syntheses of pharmaceutical intermediates or active ingredients to address some challenges often found in conventional batch processing. Typically, the issues encountered are related to safety, quality or economics and may be magnified upon scale-up. In particular, three processes will be highlighted that include: (1) controlling reaction purity profiles, (2) managing a reaction exotherm and reagent stability, and (3) converting a cryogenic synthesis to an energy-efficient non-cryogenic operation. Continuous processing approaches were implemented successfully to resolve or mitigate these process issues and enable on-scale deliveries to glass plant and pilot plant production campaigns.

134b Development of Continuous Flow Processes for Pharmaceutical Manufacturing

Jerry S. Salan, Chemical Research and Development, Pfizer, Inc., Eastern Point Road, 4016, Groton, CT 06340

The strong demand for low-cost processes in the Pharmaceutical Industry has afforded challenging opportunities for the chemical engineer. The chemical engineer toolbox is expanded when attempting to develop or convert to continuous processes. Examples of a development protocol will be discussed when assessing a process for continuous flow. The use of kinetic data will be used to determine which type of flow system is appropriate and how best to integrate the chemical reaction with the front and back ends of manufacturing processes. The discussion will include potential pitfalls and success when converting a process process from batch to continuous.

134c Assessing the Effectiveness of a Spinning Tube-In-Tube Reactor

James Ciszewski and Michael Gonzalez. U.S. Environmental Protection Agency, National Risk Management Research Laboratory, 26 W. Martin Luther King Dr., Cincinnati, OH 45268

In 2005, the American Chemical Society's Green Chemistry Institute Pharmaceutical Roundtable identified ten organic reactions extensively used in the pharmaceutical industry that are good candidates for improvements. The anticipated improvements include waste reduction, energy saving, and stereoselective optimization. The effectiveness of a spinning tube-in-tube reactor (STT®, Kreido Biofuels, Inc., Camarillo CA) to support such improvement efforts for several targeted reactions has been determined. The results will be discussed.

134d Removing Palladium from Reaction Mixtures by Fixed Bed Adsorption

Brian G. Lefebvre[1], Caitlin A. Boyd[1], Sonia M. Berberena[1], Laura E. Kuczynski[1], Pamela L. Kubinski[1], Megerle L. Scherholz[1], Dr. Stephanie Farrell[1], Robert P. Hesketh[1], and Michael J. Girgis[2]. (1) Chemical Engineering, Rowan University, College of Engineering, 201 Mullica Hill Road, Glassboro, NJ 08028-1701, (2) Chemical and Analytical Development, Novartis Pharmaceuticals Corporation, 1 Health Plaza, East Hanover, NJ 07936

The objective of this project, which was sponsored by Novartis Pharmaceuticals Corporation, was to remove palladium from a reaction mixture using a fixed bed adsorption column. Palladium is a heavy metal that is used frequently as a catalyst in many reactions in the pharmaceutical industry. Reaction mixtures can be contaminated up to 3000 ppm with palladium. A final drug substance can only contain two ppm palladium or less. For this project, the Heck coupling reaction of 4-bromobenzaldehyde and methyl acrylate was used, as it is representative of common reactions used in the pharmaceutical industry.

Adsorbents were evaluated by generating adsorption isotherms; Quadrapure TU was chosen as the adsorbent based on the high capacity of this resin at 60 °C. The adsorption column is packed with Quadrapure TU resin and glass beads. Palladium concentration is determined using a novel HPLC method that is still in development, and verified through outside analytical lab analysis. Several adsorption column runs at various flow rates and packing geometries have been performed, with the overall goal of operating the column under conditions that promote full utilization of the resin bed. The effect of feed concentration, flow rate, and packing geometry will be discussed.

134e Separation of Fine Chemical Species by Means of Continuous Chromatography: The Simulated Moving Bed Technology

Pedro Sá Gomes, Michal Zabka, Viviana Silva, and Alírio E. Rodrigues. Laboratory of Separation and Reaction Engineering, University of Porto - Faculty of Engineering, Rua Dr. Roberto Frias s/n, 4200-465 Porto, Porto, Portugal

The use of continuous chromatographic counter current operation maximizes the mass transfer driving force, providing a better utilization of the adsorbent, which may have a rather low selectivity. In addition, some typical advantages of continuous configurations include reduction of solvent consumption, increase productivity and purity and less diluted product streams, when compared to conventional batch chromatography. By means of moving bed operating mode it is possible to achieve high purity even if the resolution of the two peaks is not excellent. This is contrary to batch chromatography where high resolution is vital in order to achieve high purity. The actual circulation of the solid phase inside the column and the consequent recycle presents some technical problems, namely: equipment abrasion, mechanical erosion of adsorbent, difficulties in maintaining plug flow for the solid. This clearly limits the implementation of such technologies. In order to avoid this issue, a sequence of fixed bed columns was conceived (Broughton and Gerhold [1]) in which the solid phase doesn't move in relation to a fixed referential, but the relative movement between both fluid and solid streams is created by switching all the inlet and outlet fluid streams from time to time in the direction of the fluid flow. In the simplest operating mode, the period that a certain operating configuration prevails is called the switching time. Since the solid flow is simulated by the ports shift as described above, this technology is called Simulated Moving Bed (SMB). In the last years some "non-conventional" SMB operating modes, such as asynchronous port movement and variable inlet or outlet flow rates, were proposed providing more flexibility in the design and more efficient units.

Industrially, SMB applications can be regarded as "Old" and "New" applications associated with petrochemical and pharmaceutical/fine chemistry fields, respectively, as reviewed by Sá Gomes et al, 2006 [2]. Among the first applications of this technology (back to 1960s) is the Parex® unit (UOP) for separation of para-xylene from mixtures with the C8-isomers. In the last decade, particularly in the area of drug development, the advent of SMB has provided a high throughput, high yield, solvent efficient, safe and cost effective process option. Although it had long been established as a viable, practical, and cost-effective liquid-phase adsorptive separation technique, the pharmaceutical and biomolecule separations community did not show considerable interest in SMB technology until the mid-1990s. In 1992, Daicel Chemical Industries, Ltd. first published the resolution of optical isomers through SMB. Since that time, a considerable number of articles, patents and books on the application of SMB to pharmaceutically important compounds have been published, and already approved by FDA for some API production.

The objective of this work is to present an updated review of all SMB applications to the pharmaceutical industry, with particular emphasis on the work developed in LSRE concerning the SMB application to pharmaceutical and fine chemical separations, since the early 90's to 2007 with the design and construction of a flexible SMB unit (FlexSMB-LSRE®).

1.Broughton, D.B., C.G. Gerhold, (1961) Continuous Sorption Process Employing Fixed Bed of Sorbent and Moving Inlets and Outlets. US patent no 2 985 589

2.Sá Gomes, P.; Minceva, M.; Rodrigues A. E.; (2006) Simulated Moving Bed: Old and New, Adsorption, 12:375-392.

134f Possibility of Separating Enantiomeric Compounds Continuously with Supercritical CO2 and Chiral Ionic Liquids

Hiroshi Machida, Wataru Endo, Ryosuke Taguchi, Masaru Watanabe, Yoshiyuki Sato, and **Richard L. Smith**. Chemical Engineering, Tohoku University, 6-6-11-413, Aoba, Aramaki, Aoba-ku, Sendai, 980-8579, Japan

Continuous chiral separation processes are needed for practical-scale medical products. In separation processes for enantiomeric compounds, chiral selectivity, high purity and high throughout should be considered along with waste streams associated with the methods chosen. Ionic liquids, especially when combined with CO2 in its supercritical state, have been shown to be have some advantages over organic solvents from the viewpoint of environmental burden and their versatility in chemical processing due to the ease at which their chemical characteristics can be varied by changing the anion or cation constituent. The low volatility of ionic liquids and their lack of solubility in supercritical CO2 all enhance their appeal for use in chemical separations. Thus, the ionic liquid acts as the solvent phase for reaction or dissolution of the chemical compounds and the supercritical CO2 phase acts to extract the chemical compound that has lower affinity for the ionic liquid. Many products in the pharmaceutical industry involve the separation enantiomeric mixtures in which two basic methods exist (i) formation of a diastereomer and separation of the diastereomeric mixture by usual techniques and (ii) use of a chiral selector along with some type of phase contact system that is typically chromatography. In this work, we consider the possibility of using chiral ionic liquids as chiral selectors along with a supercritical CO2 phase. In the literature, chiral ionic liquids have been synthesized and studied, but frequently they have been found to lack chiral selectivity. On the other hand, with a CO2 phase, the chiral ionic liquid should exhibit different interactions with enantiomeric solutes and so partitioning in CO2 phase may be different between each the (R) and (S) forms. In the experiments, we use chiral ionic liquids, chiral solutes and supercritical CO2 to study the separation factors. First, properties of some ionic liquids and their characteristics with CO2 are reviewed and described. Then, a chromatographic method is used to evaluate the chiral selectivity of chiral ionic liquids. The chiral ionic liquid is used as the stationary phase in a packed column, and the CO2 was used as the mobile phase. Detection of solute was by UV/vis spectroscopy. The chiral ionic liquids studied were 1-ethyl-3-methylimidazolium-(S)-lactate [emim][(S)-lactate] and 1-butyl-3-methylimidazolium-(S)-lactate [bmim][(S)-lactate] and the racemic solutes used were 1-phenyl-1-propanol and 2-phenylpropionic acid. In some cases, only the (R)-enantiomer elutes, while in other cases, chiral selectivity can be achieved.

134g Effects of Rotation Rate, Mixing Angle, and Cohesion In Two Continuous Powder Mixers – a Statistical Approach

Patricia M. Portillo[1], Aditya U. Vanarase[1], M.G. Ierapetritou[2], and Fernando J. Muzzio[1]. (1) Chemical and Biochemical Engineering, Rutgers University, 98 Brett Road, Piscataway, NJ 08854, (2) Chemical & Biochemical Engineering Department, Rutgers University, the State University of New Jersey, 98 Brett Road, Piscataway, NJ 08854

Among emerging technologies for improving the performance of blending operations, continuous mixing currently commands enormous interest for pharmaceutical companies. Continuous processing has numerous known advantages, including reduced cost, increased capacity, facilitated scale up, mitigated segregation, and easily applied controlled shear. However, development of a continuous powder blending process requires venturing into a process that has a large and unfamiliar parametric space.

The process is as follows: as material enters the mixer, the powder crosses the pathway of several impeller blades attached to a rotating shaft placed along the axial length of the horizontal cylinder. Convective flow is the primary source of cross-sectional mixing, driven by the impeller blades. The other function of the convective motion is to axially transport the powder along the vessel (from the entrance to the discharge). The effectiveness of the mixer is particularly affected by the design of the convective system (the impeller and blades). As shown in our previous work (Portillo et al. [1]), a design parameter that affects mixing performance is the blade angle relative to the shaft.

In terms of operating conditions, the effect of impeller rotation rate, mixing angle on the powder residence time and the content uniformity of the blend exiting from two continuous powder mixers is studied. The two mixing systems examined initially in this work are both continuous blenders manufactured by GEA Buck Systems. The blenders vary in diameter and length as well as the design of the blades within the vessel. The first blender has a diameter that is 3 fold that of the second blender. Moreover, the first blender has an axial vessel that is 2.4 times longer than the second blender. For both mixers, powder samples are retrieved from the mixers outflow and characterized using Near Infrared Spectroscopy (NIR) which is commonly used to assay the content of powder samples Sánchez et al. [2].

The model blends have been formulated using Milled Acetaminophen, Lactose 100M, and Lactose 125M. The supplier (DMV International) stated that Lactose 100 M has a Hausner 1.21 ratio and Lactose 125M reported to have a 1.28 ratio (De Melkindustrie Veghel product specifications Overview).

A detailed explanation of the results for the first mixer can be found in Portillo et al. [1]. In summary we found that although increasing shear rate increased the blade passes the powder experienced, interestingly in terms of content uniformity the homogeneity worsened. Vessel inclinations resulted in longer residence times, which have also been noticed in rotary calciners (Sudah et al. [3]). Mixing performance got worse at negatively sloped inclinations, which may be a result of shorter residence times. A broader set of conditions are considered for the second mixer, including 5 mixing angles and 9 rotation rates, which were up to 300 RPM. The effect of increasing the speed further for the second continuous mixer is in good agreement with the results obtained for the first mixer.

For both blenders, 2 different Lactose powders, Lactose 100M and Lactose 125 M, a slightly more cohesive powder, are examined. The results show that the effect of powder cohesion is scale-dependent, having a significant effect in the larger mixer. However no significant effects were observed which might be due to the fact that shear rates are typically higher in convective systems than in tumblers (in fact convective blenders are often used for cohesive materials because they impart more shear), and it is possible that for both materials considered here, cohesion is simply too small to affect the outcome of the convective mixing process to a large degree.

Analysis of variance (ANOVA) is used to determine significance of main effects and their interactions. We have applied this approach to the scale-up of batch mixing where the effects of tote-size, rotation rate, fill level, and cohesion were studied (Portillo et al [4]). In this work, a randomized experimental design was used to examine the four main variables (mixer type, mixing angle, rotation rate, and powder cohesion), as well as their interactions for continuous blending. For each mixer, we initially examine a three-way ANOVA considering mixing angle, speed, and type of powder. Initially, all factors and interactions are considered, followed by a reduction in factors to remove the non-significant effects.

For the first mixer, the levels examined for the factorial design include 3 mixing angles and 2 rotation rates (16 RPM and 76 RPM), reflecting the speed limitations of the device. The data was analyzed under the assumption of normality, which was validated using a qq-plot. In terms of mixing performance, the impeller rotation rate is found to be the most significant factor, followed by the powder cohesion. The least significant factor although still significant, was found to be the mixing angle. Residence time is significantly affected by both rotation rate and mixing angle.

Applying the same procedure to study the mixing performance, the 3-way analysis was also applied to the second continuous blender. The three-way ANOVA was performed for the fractional factorial design, consisting of 9 rotation rates, 5 mixing inclinations, and 2 levels of cohesiveness. Both process parameters (rotation rate and inclination) were shown to be significant, where rotation rate is more influential than mixing angle. Cohesion on the other hand resulted in a high p-value, suggesting the insignificance of this factor. In terms of particle mobility, as expected lower shear rates and vessel inclinations resulted in longer residence times. One remaining question is whether a difference between the mixers is significant.

A 4-way ANOVA considers the results from both mixers, and reassesses the significance of other parameters when considering both datasets. The sources for the four-way statistical model are the mixer, processing inclination, rotation rate, and cohesion on the mixing performance. The ANOVA showed that cohesion was not a significant factor. Mixing angle and mixer, followed by rotation rate, are the most significant parameters. Further examining the effects of the three most significant factors, we neglect the main effect of cohesion and all interactions, the ANOVA illustrates that mixer, processing angle, and rotation rate are significant. This leads to the hypothesis that convective continuous mixers may not be significantly affected by material properties, whereas the convective design and blender geometry may be the most important operating parameters. 4-way ANOVA results from the two continuous blenders show that the blender geometry is a very significant factor.

We are expanding on the design of experiments, by examining a third continuous blender, manufactured by Gericke which is integrated with loss-in-weight feeders. The blender has diameter of 0.091 meter and length of 0.3 meter. This mixer is fixed at horizontal processing angle. The impeller has equally spaced triangular shaped 12 blades which can be arranged to propel powder forward or backward aswell as different blade angles (angle relative to the shaft). This mixer is facilitated by a weir, the weir, a semicircular disc, is fixed at the exit of the mixer. Rotating the weir angle essentially changes the powder fill level in the mixer.

The processing conditions examined in this mixer are impeller rotation rate (0-290 RPM), weir position (horizontal/rotated), and blade pattern (blade angle, forward/backward direction). The material parameters examined with respect to the powder are feed rate and concentration of API (Active Pharmaceutical Ingredient). The experimental design used is a split-split plot design. The set proposes two replications and two levels of: feed rate, API concentration, weir position, and blade configuration in addition to examining three levels of impeller rotation rate. and blade configuration form the whole plots. Weir position is randomized within each blade configuration to form a split plot, at each blade-weir replication, the remaining parameters are randomized to form a split –split plot design.

So far we have found that the residence time in the mixer and the fill level are affected by the rotation rate, weir position, and the blade angle. Homogeneity of the powder at the exit of the mixer is found to be a strong function of residence time. Interesting results so far suggest that the mean residence time does not seem to be impacted by variation of the flow rate (30 kg/hr, 55 kg/hr). A detailed experimental study to examine statistical significance of all the parameters is in progress.

References

[1] Portillo P.M., Ierapetritou M.G., Muzzio F.J., 2008, Characterization of continuous convective powder mixing processes, Powder Technology, 182, 368-379.

[2] Sánchez F.C., Toft J., van den Bogaert B., Massart D. L., Dive S.S., Hailey P., 1995, Monitoring powder blending by NIR spectroscopy, Fresenius' Journal of Analytical Chemistry, Volume 352, Numbers 7-8 / January, 1995.

[3] Sudah O.S., Chester A.W., Kowalski J.A., Beeckman J.W., Muzzio F.J., 2002b, Quantitative characterization of mixing processes in rotary calciners, Powder Technology, 126, 166-173.

[4] Portillo P.M., Ierapetritou M.G., Tomassone S., McDade C., Clancy D., Avontuur P.C., Muzzio F.J., Quality by Design Methodology for Development and Scale-Up of Batch Mixing Processes, Submitted to Journal of Pharmaceutical Innovation Feb. 2008.

The celecoxib manufacturing process was redesigned with Green Chemistry objectives as part of the project's primary goals and resulted in dramatic environmental and worker safety improvements in the manufacture of the active ingredient in the medicine Celebrex®. While most submissions from the Pharmaceutical industry to the EPA Presidential Green Chemistry Award competition involve the significant redesign of the synthetic route, the simplicity of the celecoxib molecule precluded this approach. Thus, in order to make any significant improvement in the environmental burden in the production of celecoxib, a different approach was necessary. The only viable approach was to reduce the purification requirements by reducing the amount of the major impurities formed during the synthesis. A kinetic model for the formation of these impurities was therefore developed. During these studies, the elucidation of two unprecedented reaction mechanisms responsible for the formation of an isomeric impurity was identified. Application of this new mechanistic/kinetic understanding provided a significant increase in the process efficiency with respect to raw materials, solvents, energy and waste. For example, waste was reduced by 69 percent (greater than 5000 metric tons per year at current production levels.) The regulatory challenges of developing and implementing a pharmaceutical manufacturing process that isolates the final active pharmaceutical ingredient directly from a reaction mixture will be discussed.

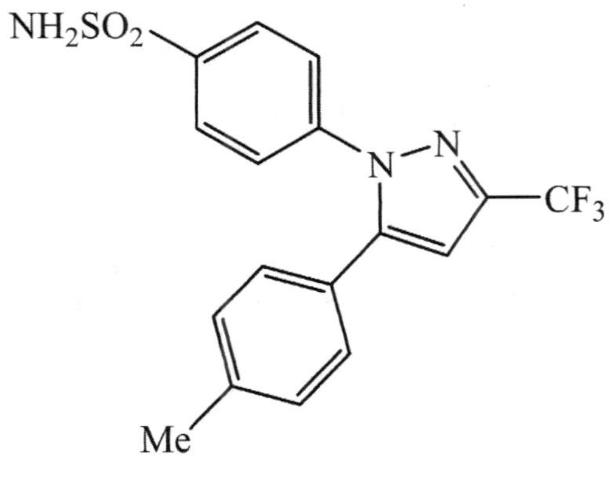

Celecoxib

223b Asymmetric Bioreduction of Nitro-Substituted Compounds with Three Enoate Reductases

Yanto Yanto, Mélanie Hall, and Andreas S. Bommarius. School of Chemical and Biomolecular Engineering, Georgia Institute of Technology, 315 Ferst Drive, Atlanta, GA 30332-0363

The use of biocatalysts in enantiomerically pure compound synthesis garners increasing attention in fine chemistry and pharma. Bioreductions of activated alkenes by enoate reductases is one of the emerging biosynthetic tools for substrates such as enals, enones, and nitroalkenes. In this study, we have characterized three enoate reductases for their reduction potential in nitro-substituted compounds. High degrees of conversion, regio- and stereo-selectivity were observed for reduction of substrates such as 1-nitro-2-phenylpropene. In order to broaden the applicability of the enzymes, we investigated the substrate specificity using wide range of C=C activating groups including aldehyde-, ketone-, imide-, and carboxylic acid- moieties. The study explored the possibilities of reduction amination of nitro-substituted alkenes for efficient synthesis of asymmetric amines. Lastly, the enoate reductases also coupled with glucose dehydrogenase for development of efficient cofactor recycling system.

223c Water Tolerant Lewis Acid Catalysis of Organic Reactions In High Temperature Water

Natalie A. Rebacz and Phillip E. Savage. Chemical Engineering, University of Michigan, 2300 Hayward Dr., Ann Arbor, MI 48109

High-temperature water (HTW), which can be loosely defined to include both liquid water above its ambient-pressure boiling point, and supercritical water, has been gaining recognition as an environmentally friendly reaction medium. At the same time, research of water-tolerant Lewis acids (WTLAs) has been steadily broadening the applications for these catalysts. Our research combines HTW with catalysis by WTLAs to achieve organic transformations in high yields. Thus far, our work has focused on alkyne hydration to form ketone, aryl-methyl ether deprotection, and Friedel-Crafts alkylation by diphenylmethanol. Alkyne hydration forms the Markovnikov product in excellent yields at 200°C with 0.5mol% $In(OTf)_3$ loading. Deprotection, a pharmaceutically important transformation, is studied with the model compound anisole which transforms to phenol at 250°C in the presence of 5 mol % $In(OTf)_3$. Diphenylmethanol successfully alkylates 1-naphthol and phenol at 200°C with 5 mol% $In(OTf)_3$. Our work focused on comparing catalysis by In(OTf)3 to that by other WTLAs, such as $Sc(OTf)_3$, $Yb(OTf)_3$, and $InCl_3$, as well as on revealing the kinetics and thermodynamics of these systems and comparing them with the conventional analog.

COMBI-CLEA FOR REACTIONS IN CONTINUOUS FLOW SUPERCRITICAL CO_2

Betti Kondor [1,2], *Juan García-Serna* [1], *Martyn Poliakoff* [2], *Neil R. Thomas* [2] *and María José Cocero* [1]

[1] *Universidad de Valladolid, Valladolid, SPAIN*

[2] *University of Nottingham, Nottingham, UK*

Abstract

The production of pharmaceuticals in a 'green' way is a concern in the recent years. In this work, we have examined an oxidation reaction in a continuous flow supercritical CO_2 system using individual CLEA (Cross Linked Enzyme Aggregate) of cholesterol oxidase and combi-CLEA of cholesterol oxidase and catalase. Besides the activity studies, the research was aimed to be extended to kinetic studies and modelling of the system to optimise the flow of H_2O_2 between the two enzymes and the production of cholestenone. The reaction examined is the oxidation of cholesterol (Scheme 1). Cholesterol oxidase (chox) catalyses the oxidation of cholesterol using O_2 to 4-cholesen-3-one, a precursor of interest in the pharmaceutical industry. H_2O_2 is produced as by-product which can be reduced to water and oxygen using the enzyme catalase. The elimination of H_2O_2 is important as it can inhibit the first reaction.

Introduction

Enzymes are catalysts with a great potential due to the fact that they usually operate at mild conditions with high chemo-, regio- and enantioselectivity. Biocatalytic reactions conducted in alternative solvents, such as supercritical CO_2 can provide chemical transformations with a lower environmental impact (1-5).

The reaction of interest in the current study is the enzymatic oxidation of cholesterol. The reaction was investigated via a combi-enzymatic reaction of cholesterol oxidase and catalase. The total output of the coupled reaction is the oxidation product (4-cholesten-3-one), water and oxygen (Scheme 1). The main product, 4-cholesten-3-one, has a fat accummulation-inhibitory effect, it is used as an active component for the prevention of obesity and precursor to contraceptive pill steroid.

The oxidation of cholesterol produces H_2O_2 as a by-product which can inhibit the oxidase. In order to prepare a one-pot enzymatic catalyst, catalase is combined with cholesterol oxidase to decompose H_2O_2 increasing the yield of cholestenone.

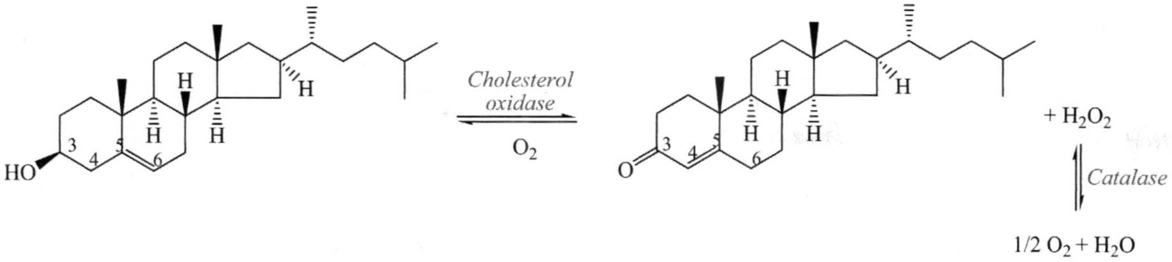

Scheme 1. Reaction pathway for the enzymatic oxidation of cholesterol.

Cholesterol oxidase immobilised on glass beads has previously been shown to be active in a continuous flow $scCO_2$ system for the oxidation of cholesterol and also for quantifying cholesterol in food samples (7-10). Both cholesterol oxidase and catalase have been previously immobilised on solid supports, however, their carrier-free immobilisation has not been previously reported. Using the CLEA technique, carrier-free biocatalyst can be prepared and therefore there is only a very small amount of inactive mass carried by the catalyst. In addition, a combi-chox-catalase CLEA has not been reported in the literature. The individual and combi-CLEAs have significant potential to be active in a continuous flow $scCO_2$ according to the success of previously reported investigations.

Carbon dioxide is a well-known green solvent (11). The main specific benefits of using supercritical CO_2 to conduct the oxidation of cholesterol are the high miscibility of $scCO_2$ with O_2 and the solubility of cholesterol in $scCO_2$. This is 40-fold greater than its solubility in water (12-14).

Experimental

Materials

All materials were purchased from common commercial sources. CO_2 (>99.95 %) and O_2 (ultrapure, >99.9 %) were purchased from Carburos Metalicos. Cholesterol (96 %) was purchased from Alfa Aesar, 4-cholesten-3-one (98 + %) was purchased from Acros and 5 cholestene-3-one (98 %) was purchased from Aldrich. Sodium azide (99+ %), sodium borohydride (>96 %) and glutaraldehyde (25 wt %) were bought from Fluka. Dextran polyaldehyde was prepared by us. Dextran from *Leuconostoc mesenteroides* (average mol wt 5M-40M, Sigma) was oxidised using sodium iodine (Montplet and Esteban SL. Barcelona). Cholesterol oxidase from *Nocardia sp.* was purchased for Merck and catalase from Bovine liver was purchased from Sigma. Isopropanol (>99 %) was purchased from Sigma.

Reaction apparatus

A schematic diagram of the reactor and its components can be seen in Figure 1. The data reported here has been obtained from a steady state condition of a continuous experiment.

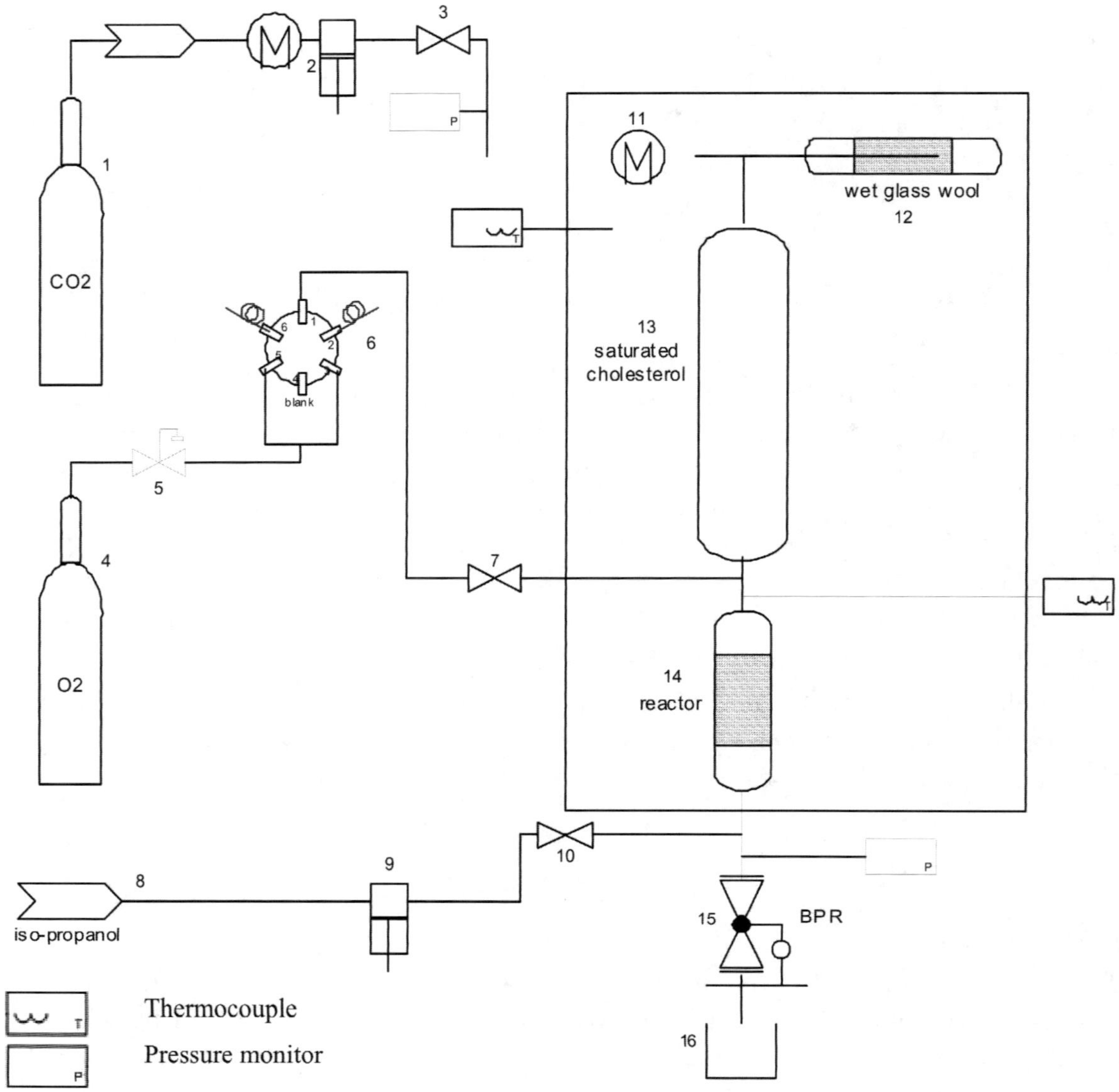

Figure 1. The small scale continuous flow reactor. Parts are labelled as follows: 1: CO_2 cylinder, 2: CO_2 cooling unit and HPLC pump, 3: Parker ball valve, 4: O_2 cylinder, 5: Manual pressure regulator, 6: Waters switching valve for O_2 dosage (two-position six-port switching valve), 7: Parker ball valve, 8: substrate/solvent reservoir, 9: HPLC pump, 10: Parker ball valve, 11: Water bath heater, 12: Vessel for CO_2 saturation with H_2O, 13: Cholesterol vessel, 14: Reactor, 15: GO valve - manual back pressure regulator (BPR), 16: Collection vial. All components are connected by 1/16" 316 stainless steel tubing. The feed streams are mixed in a heated T-piece mixer before being passed through the cholesterol vessel (14.92 cm^3 in volume) and the reactor (0.7 cm^3 in volume). The cholesterol extraction was performed up flow. Products are collected by continuous depressurisation of the reaction mixture using a manual backpressure regulator, and analysed by reverse phase HPLC.

Methods

Preparation of the cross-linked enzyme aggregates (CLEA®s)

Individual CLEAs of cholesterol oxidase (from *Nocardia sp.*) and catalase (from Bovine liver) as well as combi-CLEAs of the two enzymes were prepared. The 'CLEAtion' requires two steps, physical aggregation and cross-linking (Scheme 2). The enzymes needed no special handling. All procedures were performed at 4 °C using cold solvents.

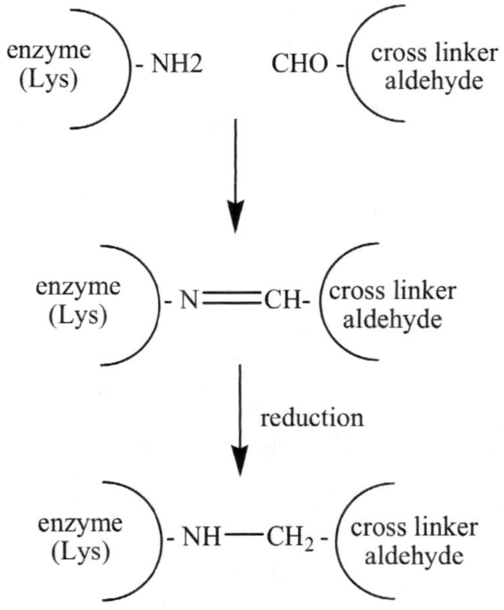

Scheme 2. Reaction of the enzyme's amino group and the cross linker's aldehyde group.

The first step is the precipitation of the enzyme that is commonly performed with organic solvents. The most common precipitants are ethanol, iso-propanol, acetone, tert-butanol, dimethoxyethane (ethyleneglycol dimethyl ether), acetonitrile and ammonium sulfate can be also used as precipitant. The precipitants (except tert-butanol) were all screened for the aggregate formation of cholesterol oxidase and catalase using 1 to 5 ratio (vol) of enzyme aliquot to precipitant. The precipitation was performed using 2.08 mg/mL cholesterol oxidase solution and 50 mg/ml catalase solution in a 5 mM sodium phosphate solution (pH 7). Cholesterol oxidase was used in low concentration due to its high price. 1800 μL cold precipitant was measured in a 2 mL centrifuge tube and 300 μL enzyme solution was added drop wise while an intense stirring. The solution was stirred for 30 minutes and centrifuged afterwards at 3000 rpm for 10 min. The supernatant was decanted and the pellet was redissolved in 1800 μL 5 mM phosphate buffer (pH 7). The enzyme activity was measured in both supernatant and the enzyme aliquot of the redissolved pellet.

The second step of CLEAtion is the cross-linking of the solid enzyme particles. The most commonly used compound is glutaraldehyde, however, other materials are also used e.g. dextran polyaldehyde, polymeric dialdehydes, paraformaldehyde. In this study, glutaraldehyde and dextran polyaldehyde were used as cross-linkers. Glutaraldehyde was used in 10 mM and 50 mM concentrations. Dextran polyaldehyde (20 % oxidised) was used in 100 μL/mL solution and 200 μl/mL solution concentrations. The cross-linking was performed in the same solution as the precipitation. The cross-linker was added to the suspension in an appropriate concentration while an intense stirring. After 18 hours stirring, the CLEAs were centrifuged at 3000 rpm for 10 min and washed two times with 5 mM phosphate buffer pH 7 (the volume of the phosphate buffer used for the washing was half of the total volume of the suspension) to remove the unreacted cross-linker. The supernatants were decanted and kept for further analysis. The washed CLEAs were resuspended in 5 mM phosphate buffer pH 7 and stored at 4 °C.

In some cases, a reduction step was also included using $NaBH_3CN$ or $NaBH_4$ as reducing agent in 0.5 mg/mL concentration. The cold borohydride solution was added to the suspension of the CLEAs (the phosphate buffer was removed prior to the addition of the solution of the borohydride solution) and stirred for 30 minutes. The CLEAs were centrifuged at 3000 rpm for 10 min and washed twice with 5 mM phosphate buffer pH 7 (the volume of the phosphate buffer used for the washing was half of the total volume). The supernatants were decanted and kept for further analysis. The CLEAs were resuspended in 5 mM phosphate buffer pH 7 and stored at 4 °C.

The combi-CLEAs were prepared using the procedure described above. The enzymes were co-precipitated.

Oxidation of cholesterol in aqueous media

The cholesterol oxidation was performed in 20 mM sodium phosphate buffer (pH 7) containing Triton X-100 surfactant in 0.5 mL/L concentration. The cholesterol solution was prepared in iso-propanol (IPA). Cholesterol oxidase from *Nocardia sp.* was prepared using 100 U (16 U/mg solid) enzyme in 3 mL 5 mM phosphate buffer (pH 7).

The reaction was carried out in a 100 mL round bottom flask at room temperature using O_2 as the oxidant. The reaction mixture consisted of 27.3 mL buffer solution, 2.5 mL cholesterol solution in IPA and 0.2 mL cholesterol oxidase. The flask was sealed with suba-seals that allowed charging the flask with O_2 using a syringe. The reaction mixture was stirred with a magnetic stirrer bar. The reaction was run up to 5 hours. The samples for the HPLC analysis (described below) were prepared by mixing 750 μl sample and 250 μl 0.2 M NaOH. The NaOH solution effectively quenched the reaction and therefore real time analysis could be performed.

Catalase assay

The catalase activity test - the reduction of H_2O_2 - was performed in 20 mM sodium phosphate buffer (pH 7) containing Triton X-100 in 0.5 mL/L concentration. H_2O_2 was used as received 30 % w/w in water. Catalase aliquot from Bovine liver was prepared in 8.33 mg/mL concentration (2950 U/mg solid) in 5 mM phosphate buffer.

The catalase assay consisted of two steps (Scheme 3). The first reaction was carried out in a 100 mL round bottom flask at room temperature using 0.2 mL enzyme suspension/aliquot. The enzyme was added to a 1 M H_2O_2 solution and incubated for exactly 1 min at room temperature while stirring. Catalase activity was quenched by inhibitor addition (sodium azide). To determine the remaining H_2O_2, a coupling reaction was performed. A dye could be formed by a reaction occurring between 4-aminoantipyrine and phenol catalysed by horseradish peroxidise (HRP). The reaction was performed in a 5 mL vial at 37 °C. The reaction mixture consisted of 0.9 mL 20 mM sodium phosphate buffer containing Triton X-100, 1 mL HRP reagent, 0.1 mL HRP solution and 3 mL sample. The HRP reagent consisted of a 6.25 mM 4-aminoantipyrine and a 100 mM phenol solutions in 1 : 1 ratio (vol). The activity of the enzyme could be calculated from the absorbance data (measured by UV-VIS spectroscopy at 500 nm) obtained for the second reaction (dye formation) of the assay. The dye formation reaction was used to determine the remaining H_2O_2 in the combi-enzymatic reaction.

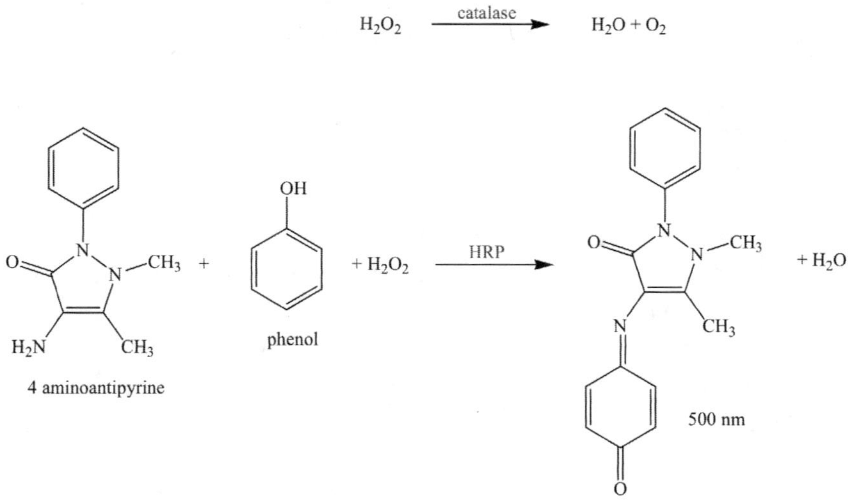

Scheme 3. Dye formation. The Dye is formed by a reaction between 4-aminoantipyrine, phenol when H_2O_2 is reduced by horseradish peroxidise (HRP).

High performance liquid chromatography (HPLC)

The oxidation of cholesterol was monitored by HPLC. The difference in polarity of cholesterol, 4-cholesten-3-one and 5-cholesten-3-one allowed the separation. A Waters HPLC instrument equipped with a Waters 717 auto sampler was used. A reverse phase C18 column was used to analyse the samples using 100 % acetonitrile as mobile phase. The column was heated to 60 °C isotherm. The flow rate of acetonitrile was 0.8 mL/min and a UV detector was used at 210 nm.

Results and Discussion

Enzyme and CLEAs activity tests

The enzyme and the CLEAs were tested before use exhibiting high activity. These results are not included in this paper but some of them will be included in the oral presentation.

Extraction of the cholesterol

Several different reactor setups were tested for the extraction of cholesterol. The reproducibility of cholesterol solubility has not been achieved in most of the cases. Presumably, the reason was the residence time. Anyhow, the objective of this study was to dissolve a quantity of cholesterol in order to perform the reaction subsequently. As the real amount of cholesterol can be know by HPLC analysis the total amount is not a critical variable in this initial stage of the reaction. This needs to be solved for performing a kinetic study.

Oxidation of the cholesterol

The reaction has been performed using an immobilised cholesterol oxidase in a carrier-free form (CLEAs) in a continuous flow of $scCO_2$ system. The oxidation reaction was examined using cholesterol oxidase (as received), cholesterol oxidase CLEAs and combi-chox-catalase CLEAs. The experiments were performed using the reactor setup shown in Figure 1.

The main results for the oxidation are presented in the tables below (Table 1-3). Cholesterol oxidase (300 U chox), cholesterol oxidase CLEA (300 U chox) and combi-chox-catalase-CLEA (200 U chox) were used as catalysts. The yield in the table was calculated as the ratio of the actual yield divided by the maximum theoretical yield in mass.

Table 1. Cholesterol oxidase 300 U chox – yield of 4-cholesten-3-one (%)

CO_2 flow (ml/min)	**40 °C 130 bar**	**40 °C 160 bar**	**50 °C 130 bar**	**50 °C 160 bar**
2 ml/min	< 0.02	< 0.5	0	0
4 ml/min	0	< 0.15	0	0
6 ml/min	< 0.2	< 0.5	0	0
8 ml/min	0	< 0.16	0	0

Table 2. Cholesterol oxidase CLEA 300 U chox – yield of 4-cholesten-3-one (%)

CO_2 flow (ml/min)	40 °C 130 bar	40 °C 160 bar	50 °C 130 bar	50 °C 160 bar
2 ml/min	< 0.2	< 1.16	0	< 0.14
4 ml/min	0	0	< 0.05	< 0.16
6 ml/min	< 0.05	< 0.05	< 0.06	< 0.11
8 ml/min	0	0	< 0.06	< 0.14

Table 3. Combi-CLEA 200 U chox - yield of 4-cholesten-3-one (%)

CO_2 flow (ml/min)	40 °C 130 bar	40 °C 160 bar	50 °C 130 bar	50 °C 160 bar
2 ml/min	< 0.04	< 0.04	na	0
4 ml/min	< 0.03	< 0.02	na	0
6 ml/min	< 0.03	< 0.02	na	0
8 ml/min	< 0.6	0	na	0

Unfortunately, as it can be seen from the results shown in the tables above that the oxidation of cholesterol in a continuous flow $scCO_2$ system exhibited a very low conversion. Although the activities of the enzyme and the CLEAs were high (measured before the use in $scCO_2$), the reaction systems could not convert cholesterol in more than ca.1%.

Cholesterol oxidase CLEA 300 U chox at 160 bar, 40 °C and 2 mL/min of CO_2 showed the highest conversion (1.16%).

Higher flow rates enhance mass transfer, but a longer reactor is required at the same time in order to keep constant residence time. This positive effect in the enzyme activity was previously reported (8) but in this case has not been demonstrated.

The enzymes may be active in flow $scCO_2$ but it is possible that they require a larger amount of water than it was provided in the experiments reported here (water-saturated CO_2 was used). A different reactor configuration needs to be tested in a further investigation.

Deactivation of the enzymes due to $scCO_2$ could be possible, but high activity was reported previously using immobilised cholesterol oxidase in a continuous flow $scCO_2$ system (7, 9). The possibilities of conformational changes under supercritical conditions and therefore deactivation of the enzyme could be possible. The effect of $scCO_2$ on the conformation of cholesterol oxidase was studied using EPR spectroscopy. There was no significant effect shown using a homogeneous enzyme (15). Here, CLEAs were examined for the reaction that might have been influenced by CO_2. Structural changes might have been caused under high pressure that could deactivate the enzyme. It is also possible that some of the active sites of the enzyme got blocked under high pressure due to compression. Further investigations needs to be performed.

Another possible cause for the low reaction rate could be the residence time (in this system the residence time in the reactor is 3.54 min at 2 mL/min CO_2 flow rate if the whole volume of the reactor is filled with catalyst). A higher conversion could be possibly obtained by providing a longer reaction time and/or providing a more active enzyme per unit. The reaction was examined using 0.5 and 1 mL/min CO_2 flow rate but the yield of 4-cholesten-3-one was still very low (the yields are not presented here but they were similar to the ones listed in the table above).

Conclusions

Individual CLEAs of cholesterol oxidase and combi-CLEAs of cholesterol oxidase and catalase were successfully prepared with high activity. Differences were seen in the activity of the cholesterol oxidase exhibited in the individual CLEAs and in the combi-CLEAs. It is more likely to be due to mass transport limitation. The large amount of catalase around the cholesterol oxidase may slow down the transport of cholesterol to the active site of the enzyme which can result in slower reaction rate. However, it is also possible that some active site of cholesterol have become blocked during the cross-linking.

The oxidation of cholesterol was performed in continuous flow $scCO_2$ using cholesterol oxidase (as received), cholesterol oxidase CLEAs and combi-chox-catalase CLEAs. Unfortunately, very low yield was obtained in all experiments performed. Further investigation needs to be performed to determine whether: 1) the enzyme needs more water, 2) CO_2 deactivates the enzyme under high pressures or 3) longer residence times are required.

References

1. Sheldon, R. A. in "Green Chemistry and Catalysis" 2007, *Wiley-VCH.*
2. Sheldon, R. A. "Selective catalytic synthesis of fine chemicals: Opportunities and trends" *J Mol Catal A: Chem*, 1996, 107, pp. 75-83.
3. Cao, L., van Langen, L., Sheldon, R.A. "Immobilised enzymes: carrier-bound or carrier-free?" *Curr Opin Biotechnol*, 2003, 14, pp. 387-394.
4. Clark, D.S. "Can immobilisation be exploited to modify enzyme activity "*Trends Biotechnol.* 1994, 12, pp. 439-444.
5. Sheldon, R. A., van Rantwijk, F. "Biocatalysis for sustainable organic synthesis" *Australian J Chem*, 2004, 57, pp. 281-289.
6. Suzuki, Kunio, "Use of 4-cholesten-3-one as an anti-obesity agent", *US Patent 5420121*, 1995.
7. Randolph, T.W. "Enzymatic Oxidation of Cholesterol Aggregates in Supercritical Carbon Dioxide" *Science*, 1988, 238, pp. 387-390.
8. Kane, M. A., Baker, G.A., Pandey, S., Bright, F.V. "Performance of cholesterol oxidase sequestered within reverse micelles formed in supercritical carbon dioxide" *Langmuir*, 2000, 16, pp. 4901-4905.
9. Randolph, T. W. "Enyzme-catalysed oxidation of choelsterol in supercritical carbon-dioxide" *AIChE*, 1988, 34, pp. 1354-1360.

10. Bauza, R., Ríos, A., Valcárcel, M. "Coupling immobilized enzymes flow reactors with supercritical fluid extraction for analytical purposes" *Analyst*, 2002, 127, pp. 241-247.

11. Jessop, P.G., Leitner, W. "Chemical synthesis using supercritical fluids" *Wiley-VCH*, 1999.

12. Jimmy Yun, S.L., Keat Liong, K., Gurdial, G.S., Foster, N.R. "Solubility of Cholesterol in Supercritical Carbon Dioxide", *Ind. Eng. Chem. Res.*, 1991, 30, pp. 2476-2482.

13. Singh, H., Yun, J.S., Macnaughton, S.J., Tomasko, D.L., Foster, N.R. "Solubility of cholesterol in supercritical ethane and binary gas mixtures containing ethane", *Ind. Eng. Chem. Res., 1993,* 32, pp. 2841-2848.

14. Saad, H.S., Higuchi, W.I. "Water solubility of cholesterol" *J. Pharmaceutical Sci.*, 1965, pp. 1205-1206.

15. Randolph, T.W., Clark, D.S., Blanch, H.W., Prausnitz, J.M. "Cholesterol aggregation and interaction with cholesterol oxidase in supercritical carbon dioxide" *Biophysics*, 1988, 85, pp. 2979-2983.

KINETICS OF HYDROFORMYLATION OF HIGHER OLEFINS USING RHODIUM-PHOSPHITE CATALYST IN A THERMOMORPHIC SOLVENT SYSTEM

Maizatul S. Shaharun, Hilmi Mukhtar, Suzana Yusup and Binay K. Dutta, Universiti Teknologi PETRONAS, Malaysia*

Abstract

The liquid-liquid biphasic thermomorphic or temperature dependent multicomponent solvent (TMS) is an attractive alternative to the conventional single or biphasic reaction media system. In such a solvent, the catalyst remains as a residue in one of the liquid phases and the product goes preferably to the other liquid phase providing ease of product separation and catalyst recovery. This paper describes the kinetics of hydroformylation of two higher olefins (1-octene and 1-dodecene) using a homogeneous catalyst consisting of $HRh(PPh_3)_3(CO)$ and $P(OPh)_3$ in a TMS- system containing propylene carbonate (PC), dodecane and 1,4-dioxane. The effect of concentration of the olefins, catalyst loading, partial pressure of CO and H_2 and temperature on the rate of reaction has been studied at temperature of 353, 363 and 373 K. The rate was found to be first order with respect to catalyst, 1-octene and partial pressure of H_2. The rate vs. P_{CO} shows a typical case of substrate inhibited kinetics. Three generalized rate models were developed on the basis of plausible reaction pathways. The kinetic and equilibrium parameters of the models were estimated by nonlinear least square regression of experimental data. Conventional model discrimination techniques were used to identify the correct one. The rate model with H_2-addition as the controlling step was found to represent the rate data. The rate model predicted the conversion of the two alkenes satisfactorily with an average absolute error of ±4.0 %.
Keywords: Hydroformylation; Thermomorphic solvent systems; Kinetics; 1-Octene; 1-Dodecene; Rhodium catalyst.

Introduction

Hydroformylation is an attractive process of synthesis of linear aldehydes, the linear ones in particular. These aldehydes are the starting materials for the synthesis of linear alcohols.[1] The overall reaction can be represented by Eq. (1).

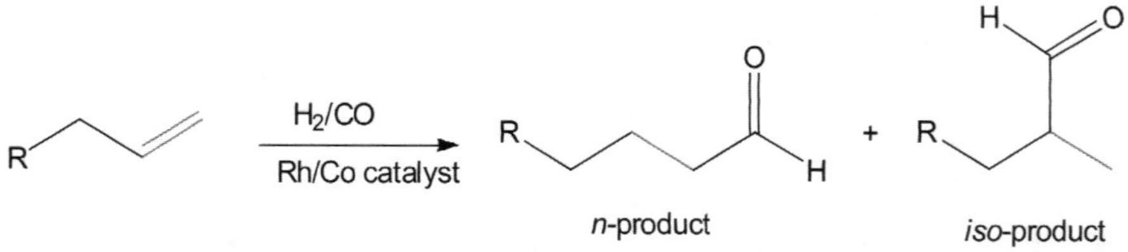

(1)

Different types of catalysts and solvents have been developed over the past years for this industrially important catalytic process. A rhodium complex with phosphine or phosphite ligand is a recent development in hydroformylation catalysts. The catalyst is expensive and its recovery and reuse are the major factors to determine its commercial competitiveness. Development of the aqueous biphasic solvent system has been a major milestone. The water soluble catalyst preferentially remains in the aqueous phase facilitating its recovery at the end of the reaction. The reaction occurs principally in the hydrocarbon phase of this biphasic solvent mixture. However, the process was not very attractive for synthesis of higher aldehydes.[2] This limitation led to the development of a novel solvent system – the TMS system or temperature dependent multicomponent solvent system. The TMS-system provides easy separation of the expensive catalyst from the products on cooling of the reaction mixture.[3,4,5] The isomerizing hydroformylation of trans-4-octene in the TMS-system of propylene carbonate (PC)/dodecane/*p*-xylene has been carried out producing a very high conversion (about 99%) of the trans-4-octene and offering very attractive selectivities of *n*-nonanal ranging from 79 to 90%[6] but a strong rhodium leaching as high as 47% was reported. On the other hand, Tijani et al.[6] has developed a thermomorphic biphasic rhodium system using an inexpensive and conventional ligand such as $P(OPh)_3$ to catalyze the hydroformylation of higher olefins (>C6). However, neither experimental data nor any empirical model on the kinetics of the hydroformylation of higher alkenes in the TMS-system are available in the literature.

Here we report the kinetics of hydroformylation of 1-octene and 1-dodecene in the TMS-systems consisting of PC, dodecane and 1,4-dioxane. In the above context, the specific objectives of this work are to: (1) study the kinetics of the hydroformylation of 1-octene and 1-dodecene with the above catalyst and reaction medium, (2) develop a mechanistic rate model of the reaction and evaluate the rate parameters by fitting experimental data, and (3) assess the validity and quality of the developed rate model by fitting experimental hydroformylation data.

Experimental

Materials

1-Octene, 1-dodecene, triphenylphosphite, propylene carbonate, dodecane and 1,4-dioxane, purchased from Merck, were 97-99 % pure. The chemicals were used without further purification. No impurity could be detected by GC analysis. Nitrogen and syngas (1:1 CO/H_2) were supplied by MOX, with purity of 99.99 %.

Experimental setup and procedure

The catalyst was prepared in situ by mixing a rhodium catalyst precursor, $HRh(CO)(PPh_3)_3$ (ABCR, Germany) and triphenylphosphite, $P(OPh)_3$ in molar ratio of 1:12. Experimental runs were conducted at different temperatures, partial pressure of H_2 and CO and catalyst concentrations varying from 1.8×10^{-4} to 8.3×10^{-4} $kmol m^{-3}$ (0.17 to 0.76 g L^{-1}). Hydroformylation of 1-octene was carried out in a stirred high pressure reactor (model: Parr 4843) equipped with a stirrer and temperature control system. The feed (240 mL) mixed with the catalyst was added to the preheated reactor. The reactor was purged with nitrogen and syngas successively, and then adjusted to the desired temperature. It was pressurized with an equimolar mixture of CO and H_2 to a desired total pressure which was maintained constant during the whole run. The stirrer speed was set at 450 rpm while the reaction continued. Liquid samples (each less than 1 mL) were withdrawn at regular time intervals to follow the progress of the reaction. The reaction was stopped quickly by cooling down the mixture to the room temperature. The gas was vented out and the biphasic mixture was separated. The analysis of reactants and products was carried out by a gas chromatographic method using 5 % phenylmethyl siloxane capillary column. Each type of experiment was repeated three times to check for reproducibility. Measurements are, in general, reproducible within a maximum of 10 % but often within a few per cent.

Solubilility measurement

The solubilities of pure H_2 and CO in TMS-systems of PC/dodecene/1,4-dioxane were measured by the absorption method at different pressures (1-25 bar) and temperatures (353-373 K). Requisite volume of the solvent only was taken in the reactor which was then quickly pressurized. The contents were stirred at 450 rpm for about 1 hr to equilibrate the liquid phase with the gas. A few preliminary experiments showed that this time was sufficient for equilibriation. The total pressure change due to absorption was recorded on-line as a function of time till it attained a steady and constant value, indicating saturation of the liquid phase. From the initial and final pressure readings, the solubility was calculated in kmolm^{-3} as:

$$C_L = \frac{\left[P_i - (P_f - p^v)\right]V_g}{RTV_L} \qquad (2)$$

where C_L represents the concentration of the solute gas in the liquid phase, P_i and P_f are the initial and final pressure readings in the reactor; p^v is the vapour pressure of the solvent; V_g and V_L are the volumes of the gas and liquid phases, respectively. Within the temperature range of this study, PC and dodecane have negligible vapour pressure. The vapour pressure of pure 1,4-dioxane was calculated with the Antoine equation.[7] In order to ensure the applicability of the ideal gas law in Eq. 2, the H_2 and CO fugacity coefficients, were calculated using the Peng-Robinson equation of state (PR EoS). The fugacity coefficients are very close to unity, and the vapour phase could be considered as an ideal gas.[8] Measurements are reproducible within 2-3 %.

Results and Discussions

Solubility of CO and H_2

For the kinetic study, knowledge on the solubility of gaseous reactants in the reaction medium is essential. The solubility of H_2 and CO in the TMS system is linearly dependent on the pressure in the pressure range of 0.1 to 2.5 MPa. Henry's law constants for solubilities of CO and H_2 in TMS system of PC/dodecane/1,4-dioxane with composition of 0.30/0.10/0.60 are presented in Table 1. These values were used in the calculation of the concentrations of dissolved CO and H_2 in the reaction medium. The effect of temperature (353 to 373 K) and pressure (0.1-2.5 MPa) on the solubility was also determined.

Kinetics

The reaction proceeds via catalytic addition of H and formyl (CHO) groups across the double bond of the olefin to give aldehydes. The main steps in the catalytic cycle follow the Heck and Breslow mechanism developed for the cobalt-catalyzed oxo reaction.[9] The initial rates of hydroformylation were determined under the range of conditions as shown in Table 2.

Table 1: Solubility of H_2 and CO in TMS-system of PC/dodecane/1,4-dioxane (0.30/0.10/0.60).

Temperature	Henry's law constant (m^3 MPa kmol^{-1})	
(K)	H_2	CO
353	10.13	8.05
363	9.98	7.88
373	9.79	7.56

To ensure the reaction is in the chemical reaction controlling region, preliminary experiments were carried out at temperature of 363 K with different stirring speeds. It was observed that the reaction rate was enhanced greatly with the increase of the stirring speed at first and then approached to a constant, above 400 rpm. Therefore, all the subsequent experimental runs were conducted at the fixed agitation speed of 450 rpm. This conclusion is further supported upon comparing the rates with the maximum rates of mass transfer ($\frac{R_0}{k_L a C^*_{H_2}}, \frac{R_0}{k_L a C^*_{CO}}$) of CO and H_2 at the respective conditions. Here, R_o represents the initial rate of hydroformylation, in $kmolm^{-3}s^{-1}$; k_La is the volumetric mass transfer coefficient, in s^{-1}; and $C^*_{H_2}$ and C^*_{CO} are the solubilities of H_2 and CO at the gas-liquid interphase, in $kmolm^{-3}$, respectively. The calculated ratio of initial reaction rate to the maximum mass transfer rate of gas-to-liquid was found to be less than 0.10 at 450 rpm. This clearly shows that the rates observed in the entire range of conditions studied are much smaller than the mass transfer rates that would prevail under these conditions, and hence the rate data are in the kinetic regime.

The kinetics of the reaction has been investigated at three temperatures - 353, 363 and 373 K. The rate was found to decrease with higher P_{CO} (or liquid phase concentration of CO) in the pressure range of 1.0 to 11.5 atm (Fig. 1). The rate was found to be first order with respect to the concentrations of catalyst for both the substrates, the substrate concentration as well as the partial pressure of hydrogen gas. In fact, the parametric effects were found to be similar to that of kinetics in homogeneous system under the industrial operating conditions.[9] The aldehyde concentration versus time data for each experiment were fitted by a polynomial equation which was then differentiated to determine the slope (experimental reaction rate in $kmolm^{-3}s^{-1}$) at the initial time. This rate versus concentration data were then used to determine the parameters by fitting the experimental rate.

The pathways of the hydroformylation reaction based on the Heck and Breslow mechanism are shown in Figure 3 and mechanistic rate equations were developed on this basis using the conventional quasi-steady state formalism. Qualitative reasoning and some available experimental observations form the basis of the scheme.[9,10,11,12,13] In developing the rate equation, it is assumed that (i) all but one of the reaction steps are in equilibrium, (ii) the starting intermediate species at the reaction interface is the complex $HRh(CO)_2(L)_2$., and (iii) the total rhodium concentration remains constant. Three rate equations have been developed corresponding to the three possible transition states reported in open literature.[10] These are

Table 2: Range of variables studied in the present work

	1-octene	1-dodecene
Conc. of catalyst ($kmolm^{-3}$)	$8.66\times10^{-5} – 6.78\times10^{-4}$	$8.66\times10^{-5} – 3.46\times10^{-4}$
Conc. of 1-olefins ($kmolm^{-3}$)	0.21-4.2	0.20-3.0
P_H (MPa)	0.3-1.5	0.3-1.5
P_{CO} (MPa)	0.3-1.5	0.1-1.5
T (K)	353-373	353-373
Reaction volume (mL)	240	220
Solvent	PC/dodecane/1,4-dioxane (0.30/0.10/0.60)	PC/dodecane/1,4-dioxane (0.30/0.10/0.60)

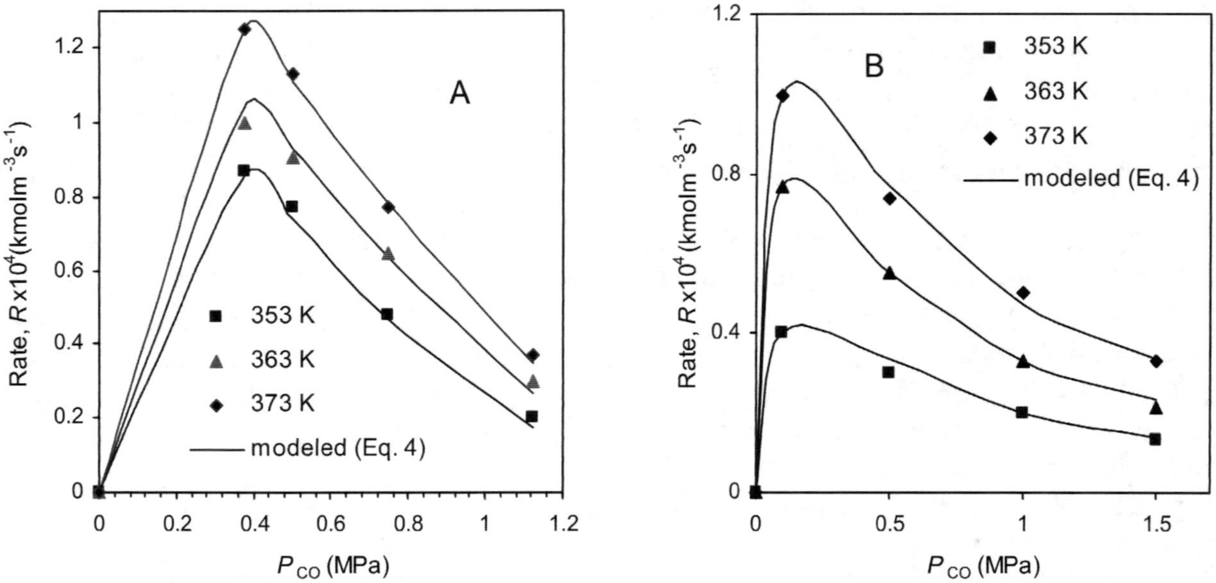

Figure 1. Effect of partial pressure of CO on the rate of hydroformylation of 1-octene (A) and 1-dodecene (B). Reaction conditions: P = 1.5 MPa, 1-olefins = 1.9 kmolm^{-3}, HRh(CO)(PPh$_3$)$_3$ = 1.73x10^{-4} kmolm^{-3}, P(OPh)$_3$/HRh(CO)(PPh$_3$)$_3$ = 12, PC/dodecane/1,4-dioxane=0.30/0.10/0.60.

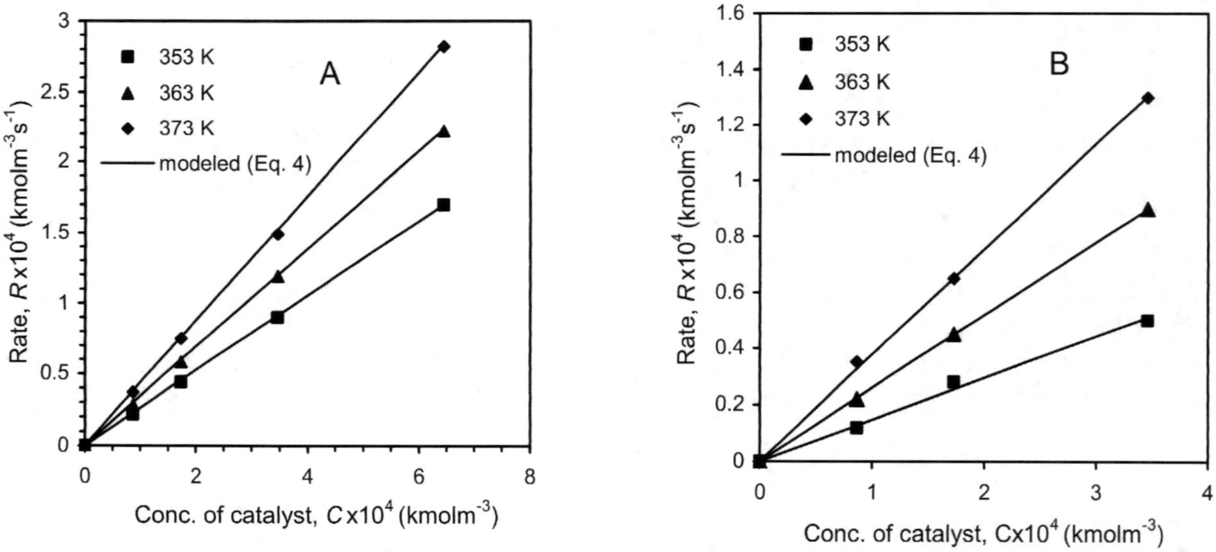

Figure 2. Effect of concentration of catalyst on the rate of reaction: 1-octene (A) and 1-dodecene (B). Reaction conditions: P = 1.5 MPa, 1-octene = 1.9 kmolm^{-3}, P(OPh)$_3$/HRh(CO)(PPh$_3$)$_3$ = 12, PC/dodecane/1,4-dioxane=0.30/0.10/0.60.

Rate Model M1: Oxidative addition of H_2 is rate –controlling

$$\text{Rate} = \frac{k[\text{catalyst}][\text{CO}][\text{alkene}][\text{H}_2]}{1 + K_1^*[\text{CO}] + K_2^*[\text{CO}]^2 + K_3^*[\text{alkene}] + K_4^*[\text{CO}][\text{alkene}]} \quad (3)$$

where k is the reaction rate constant of the rate determining step and K_1^* - K_5^* are 'equilibrium constants' of the other reaction steps in the pathways shown in Figure 3.

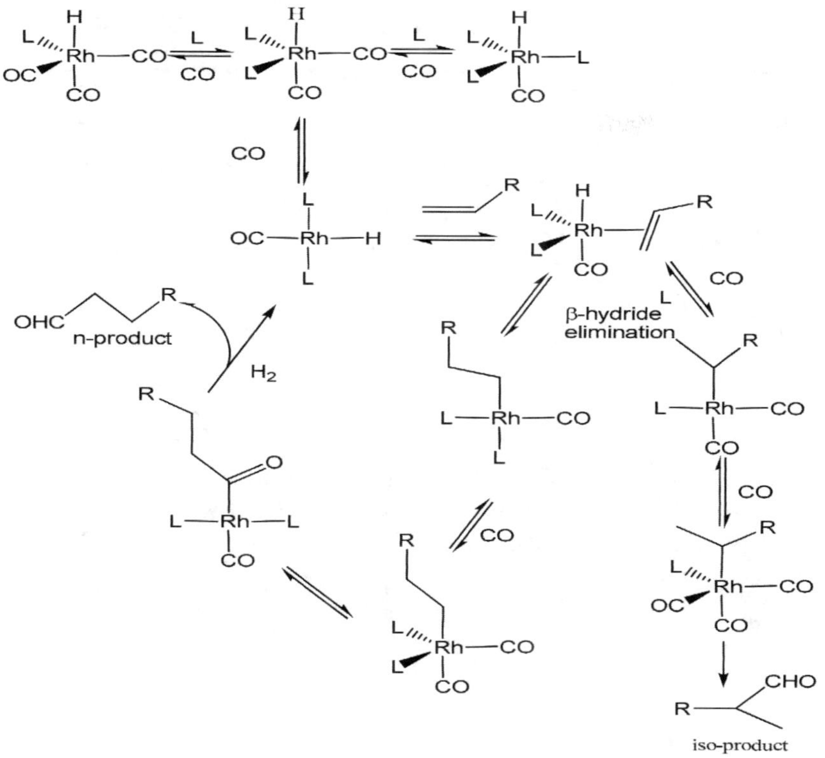

Figure 3: Plausible steps for rhodium catalyzed hydroformylation in the presence of excess concentration of $P(OPh)_3$ ligand.

Rate Model M2: The insertion of the alkene into Rh-complex is rate-controlling

$$\text{Rate} = \frac{k[\text{catalyst}][\text{CO}][\text{alkene}][\text{H}_2]}{\left(1 + K_1^*[\text{CO}]^2 + K_2^*[\text{alkene}] + K_3^*[\text{CO}][\text{alkene}]\right)[\text{H}_2] + K_4^*[\text{CO}][\text{alkene}]} \quad (4)$$

and

Rate Model M3: Formation of the acyl complex is rate-controlling:

$$\text{Rate} = \frac{k[\text{catalyst}][\text{CO}][\text{alkene}][\text{H}_2]}{\left(1 + K_1^*[\text{CO}] + K_2^*[\text{CO}]^2 + K_3^*[\text{alkene}] + K_4^*[\text{CO}][\text{alkene}]\right)[\text{H}_2] + K_5^*[\text{CO}][\text{alkene}]} \quad (5)$$

Non-linear least square regression based on the criterion of minimization of the mean residual sum of squares (MRSS) was performed to determine the kinetic parameters where N_{expt} is the number of experimental data, N_{param} is number of model parameters, R_{calc} and R_{expt} represent calculated and experimental rates, respectively.

$$MRSS = \frac{\sum_{i=1}^{N} \left(R_{calc} - R_{\text{expt}}\right)^2}{N_{\text{expt}} - N_{param}} \quad (6)$$

The activation energy of the rate–controlling reaction step and the pre-exponential factor were obtained from the Arrhenius equation with 'temperature centering'.

$$k_i = A_i \exp\left[\frac{-E_i}{R}\left(\frac{1}{T} - \frac{1}{T_m}\right)\right] \quad (7)$$

Table 3: Estimated kinetic model parameters with 95% confidence limits

Substrate	T (K)	k	K_1^*	K_2^*	K_3^*	K_4^*	SEE	E_a (kJ mol^{-1})
1-octene	353	9.6×10^2	9.1×10^{-3}	4.4×10^2	9.3×10^{-3}	3.32	4.5	66.3
C_8	363	1.6×10^3	20.3	3.5×10^2	1.2×10^{-1}	8.8×10^{-1}		
	373	3.1×10^3	55.6	3.3×10^2	8.5×10^{-1}	6.4×10^{-7}		
1-dodecene	353	7.1×10^2	0.37	2.6×10^3	0.86	54.1	3.6	76.1
C_{12}	363	1.3×10^3	42.4	2.9×10^3	0.45	42.1		
	373	2.8×10^3	139.0	3.2×10^3	7.2×10^{-2}	23.9		

Here T_m is introduced for temperature centering. In the present study T_m = 363 K was used for temperature centering and parameter estimation. This scaling procedure admits of robust parameter estimation minimizing non-linear correlation effects.[14,15,16] An 'equilibrium constant', K, was similarly determined as a function of temperature using the van't Hoff equation also with temperature centering.

$$K_i^* = \exp\left[\frac{\Delta S_i}{R} - \frac{\Delta H_i}{R}\left(\frac{1}{T} - \frac{1}{T_m}\right)\right] \quad (8)$$

The error estimate given by Eq. (6) together with thermodynamic consistency of the calculated activation energy or negative kinetic parameter, if any, were used as the criteria of discrimination among the three rate models.[16,17,18] Model M2 was rejected because it displayed a non-linear Arrhenius plot. Model M3 showed high average standard error estimate ranging from 11% to 18% besides having negative kinetic parameters. The rate model M1 is able to describe the experimental data reasonably well over the whole range of pressure, concentrations of the olefins (1-octene and 1-dodecene) and temperature.

As Figure 4 displayed convincingly, the rate model M1 describes the experimental data of both substrates well. The estimated parameters as well as the activation energy and error

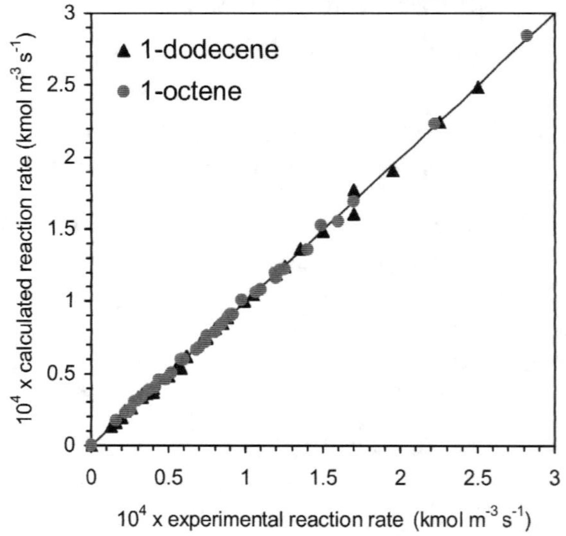

Figure 4. Parity plot of the model predictions of the rate of hydroformylation of 1-octene and 1-dodecene in TMS-systems of PC/dodecane/1,4-dioxane (0.30/0.10/0.60).

estimates are presented in Table 3. These results suggest that a mechanism of reaction featuring oxidative addition of H_2 to acylrhodium intermediate species as rate determining is fair to describe the hydroformylation of the two substrates – namely, 1-dodecene and 1-octene. The average deviation between experimental and calculated rate of reaction was found to be in the range of ± 4.0 % and the activation energy of 66.3 and 76.1 $kJmol^{-1}$ were determined for the hydroformylation of 1-octene and 1- dodecene, respectively. The results also support previous findings in demonstrating that under the experimental conditions listed in Table 1, the acylrhodium complex is the resting state and oxidative addition of H_2 is rate determining.[19]

Conclusion

The kinetics of hydroformylation of 1-octene and 1-dodecene a homogeneous catalyst consisting of $HRh(PPh_3)_3(CO)$ and $P(OPh)_3$ in a TMS- system containing propylene carbonate (PC), dodecane and 1,4-dioxane has been investigated in a temperature range of 353-373 K. The rate versus P_{CO} shows a typical case of substrate inhibited kinetics. The rate was found to be first order with respect to concentrations of the catalyst, the olefins and hydrogen in the reaction mixture. A mechanistic rate equation has been proposed based on plausible mechanism steps. It was possible to fit reasonably well the experimental data considering oxidative addition of hydrogen as rate-determining. The validity and quality of the developed rate model was assessed with respect to C_{12} substrate (1-dodecene). The rate model predicted the conversion of the 1-octene and 1-dodecene satisfactorily with maximum deviation of 4.5 %.

Notation

[catalyst]	concentration of catalyst $HRh(CO)(PPh_3)_3/P(OPh)_3$ in the reaction mixture, kmol m^{-3}
[CO]	concentration of carbon monoxide in the reaction mixture, kmol m^{-3}
$[H_2]$	concentration of hydrogen in the reaction mixture, kmol m^{-3}
[alkene]	concentration of alkene in the reaction mixture, kmol m^{-3}
E	activation energy for rate constant, kJ mol^{-1}
ΔH	heat of reaction, kJ mol^{-1}
k	reaction rate constant, units will be specific to the form of the rate expression
A	pre-exponential factor, units will be identical to the rate constant
R	gas constant, 8.314 J mol^{-1} K^{-1}
ΔS	entropy of reaction, J mol^{-1} K^{-1}
T	temperature, K

Acknowledgement

Financial assistance from Ministry of Science,Technology and Innovation (MOSTI),Government of Malaysia, under the Research Project No.03-02-02-SF0019:*Development of a green process for the production of higher aldehydes from olefins by hydroformylation* is gratefully acknowledged.

References

1. Bohnen H, Cornils B. Hydroformylation of Alkenes: An Industrial View of the Status and Importance. ChemInform. 2003; 34:17.

2. Beller M, Cornils B, Frohning CD, Kohlpaintner CW. Progress in hydroformylation and carbonylation. J. Mol. Catal. A, 1995; 104:17-85.
3. Behr A, Obst D, Turkowski B. Isomerizing hydroformylation of trans-4-octene to n-nonanal in multiphase systems: acceleration effect of propylene carbonate. J. Mol. Catal. A: Chem. 2005; 226:215-219.
4. Behr A, Roll R. Hydroaminomethylation in thermomorphic solvent systems. J. Mol. Catal. A: Chem. 2005; 239:180-184.
5. Behr A, Fangewisch C. Rhodium-catalysed synthesis of branched fatty compounds in temperature-dependent solvent systems. J. Mol. Catal. A: Chem. 2003; 197:115-126.
6. Tijani J, Ali BE. Selective thermomorphic biphasic hydroformylation of higher olefins catalyzed by $HRhCO(PPh_3)_3/P(OPh)_3$. Applied Catal. A: General 2006; 303:158-165.
7. Romero C, Villares A, Haro M, Giner B, Lafuente C. Experimental and predicted vapor-liquid equilibrium of 1,4-dioxane with cycloalkanes and benzene. Fluid Phase Equilibria, 2005; 238:1-6.
8. Peng DY, Robinson DB. A new two parameters Equation of State. Ind. Eng. Chem. Fundam. 1976; 15:59-64.
9. van Leeuwen, P.W.N.M. Homogeneous catalysis, understanding the art. Kluwer Academic Publishers, Dordrecht, The Netherlands. Springer-Verlag, 2004.
10. Musaev DG, Matsubara T, Mebel AM, Koga N, Morokuma K. Ab initio molecular orbial studies of elementary reactions and homogeneous catalytic cycles with organometallic compounds. Pure & Appl. Chem. 1995; 67:257-263.
11. Brown JM, Kent AG. Structural characterisation in solution of intermediates in rhodium-catalysed hydroformylation and their interconversion pathways. J. Chem. Soc., Perkin Trans. 1987; 2:1597-1607.
12. Jongsma T, Challa G, van Leeuwen PWNM. A mechanistic study of rhodium tri(*o*-t-butylphenyl)phosphite complexes as hydroformylation catalysts. Organometal. Chem. 1991; 421:121.
13. van der Veen LA, Keeven PH, Schoemaker GC, Reek JNH, Kramer PCJ, van Leeuwen PWNM., Lutz M, Spek AL. Origin of the Bite Angle Effect on Rhodium Diphosphine Catalyzed Hydroformylation. Organometallics, 2000; 19:872.
14. Pant KK, Kunzru D. Catalytic pyrolysis of n-heptane: Kinetic and modeling. Industrial & Engineering Chemistry Research. 1997; 36:2059-2065.
15. Wojciechowski BW, Rice, NM. Experimental methods in kinetic studies. Elsevier Science BV., Amsterdam, 2003.
16. Patel S, Pant KK. Experimental study and mechanistic kinetic modeling for selective production of hydrogen via catalytic steam reforming of methanol. Chemical Eng. Science, 2007; 62:5425-5435.
17. Pengpanich S, Meeyoo V, Rirksomboon T, Bunyakiat K. Catalytic oxidation of methane over CeO_2-ZrO_2 mixed oxide solid solution catalysts prepared via urea hydrolysis. Applied Catalysis A: General 2002; 234:221-233.
18. Hurtado P, Ordonez S, Sastre H, Diez FV. Development of a kinetic model for the oxidation of methane over Pd/Al_2O_3 at dry and wet conditions. Applied Catalysis B: Environmental. 2004; 51:229-238.
19. van Leeuwen PWNM, Claver, Carman. Rhodium Catalyzed Hydroformylation. Kluwer Academic Publishers, Dordrecht, The Netherlands. Springer-Verlag, 2000. http://www.knovel.com/knovel2/Toc.jsp

290a Green Separations in the Pharmaceutical Industry: Supercritical Fluid Extraction and Chromatography

Ram B. Gupta, Chemical Engineering, Auburn University, Auburn, AL 36849-5127

The key attributes of a green separation are energy efficiency and low environmental impact providing a sustainable operation. On a unit mass basis, pharmaceutical industry carries a significant burden of separation due to the need for high purity products. Due to environmentally benign nature of CO_2, supercritical fluid extraction (SFE) and supercritical fluid chromatography (SFC) are making inroads in pharmaceutical separations. The high equipment cost is being offset by savings in the energy cost and improvement in the product quality. The key applications are in the control of process impurities such as chemical intermediates and residual solvents, and in polymorphic control and chiral resolution. From supercritical fluid, the product can be recovered by simple depressurization. This instant solute-supercritical fluid separation is a specific feature of SFC, as opposed to that in liquid chromatography where the solute recovery and solvent regeneration is done in a separate downstream process. In addition, due to the low viscosity combined with high diffusivity, SFC is about 3-5 times faster than liquid chromatography. A notable example is the much improved chiral separation in the pharmaceutical industry. The use of CO_2 in separations does not add the greenhouse effect. CO_2 comes as a byproduct from the chemical industry or beverage fermentation. In fact, due to the reduction in energy use (as compared to solvent based processes), SFC and SFC help reduce the net CO_2 emission.

Analysis of Separation Methods for Isopropanol Recovery in the Celecoxib Process

C. Stewart Slater[1], Mariano Savelski[1],
Gregory Hounsell[2], Daniel Pilipauskas[2], Frank Urbanski[2]
(1) Rowan University, Glassboro, NJ
(2) Pfizer, Inc., New York, NY

Various approaches to solvent recovery have been studied for the manufacturing process of celecoxib, the active ingredient in Celebrex®. A design case study has been performed by Rowan University with Pfizer through a green engineering partnership program. The manufacturing operation at Barceloneta, Puerto Rico was evaluated and several green engineering alternatives for the purification and recovery of isopropanol from waste streams proposed. This separation is complex due to the multiple waste streams generated, with varying compositions of isopropanol, ethanol, methanol, water and dissolved solids. Overall goals were waste minimization and isopropanol recovery and purification. The group performed a conceptual study of distillation, extraction, reactive distillation, adsorption, and membrane-based processes. Through use of computer simulation and literature/design methodologies, traditional methods were shown to be unable to obtain high isopropanol purities with the equipment available. Both molecular sieve adsorption and membrane pervaporation appear to have the most promise to effectively recover and purify isopropanol. Several green design approaches were evaluated using distillation combined with either molecular sieve adsorption or membrane pervaporation. The case study describes equipment and processing issues, and estimates environmental impacts and costs.

INTRODUCTION

Rowan University and Pfizer have partnered through our Engineering Clinics program to examine green engineering approaches to solvent minimization in the celecoxib process[1,2,3]. This project is one of several academic industrial partnerships with pharmaceutical and specialty chemical companies that have been conducted at Rowan University [4]. These projects are funded by industry through the Rowan Engineering Clinic program and by the U.S. Environmental Protection Agency Pollution Prevention grant program. The projects involve a process case study or problem that has a green chemistry/green engineering component. Depending on the nature of the projects, they can be "paper" projects that are design based or experimentally-based. This particular project is funded through the Pfizer Green Chemistry Program.

The project objectives were the investigation of solvent recovery alternatives to minimize waste from the celecoxib process. Celecoxib is the active ingredient in the top-selling arthritis pain medicine, Celebrex®. The objectives of the student project team were to compare the current process route with green engineering options.

- Waste stream reduction and isopropanol (IPA) recovery
- Define operational sequences
- Equipment and process steps required

- Estimate costs and environmental impacts
- Make proposals and recommendations

Students reviewed documentation on the "history" of the celecoxib process and the various process chemistries. Some of the previously suggested approaches for solvent recovery were also analyzed. A review of the capabilities of the existing equipment at Barceloneta to recover IPA from the waste was done. It was desired to work with the current capital assets, but several alternative processes involving new equipment were evaluated.

RESULTS AND DISCUSSION

The first step in the process was to analyze the waste flows per batch of celecoxib. Figure 1 shows the waste streams of interest in the minimization and recovery. This is a very efficient process with an E-factor of 9.0, but since the annual production is large, solvent waste reduction and recovery represent a significant opportunity. The study compared waste minimization options for combined waste composed of the three streams and a segregated approach. The segregated approach take streams with similar properties (centrifuge wash and dryer distillates) and combines for separation/recovery and the mother liquor (filtrate) stream is either pre-concentrated for sale or incinerated. This approach reduces the separation burden on the existing equipment inventory in Barceloneta and eliminates sending the high TDS mother liquor stream through the proposed separation process train. Based on batch flows, yearly production estimates and equipment availability, an equipment utilization schedule was developed for each process alternative described.

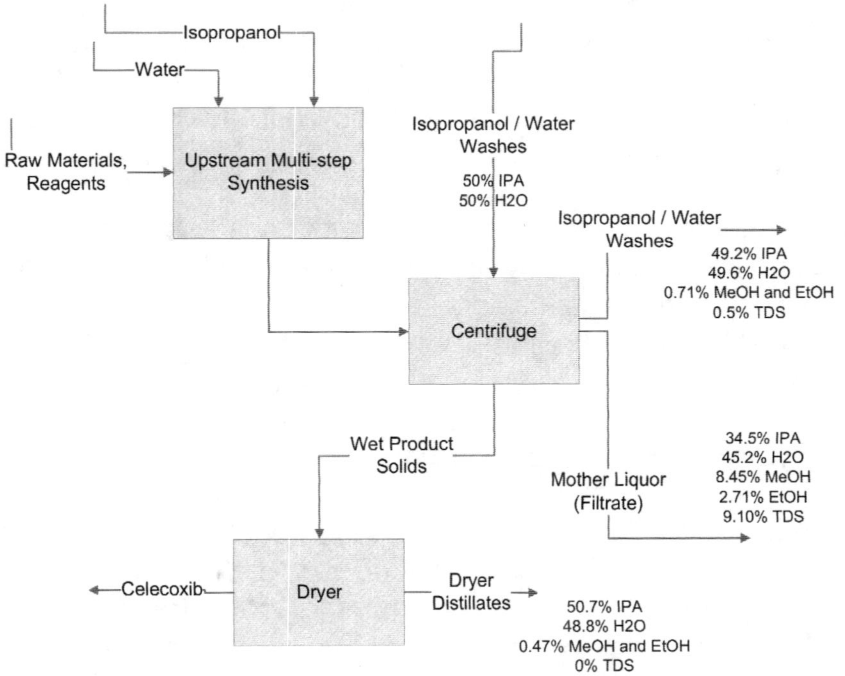

Figure 1. General celecoxib process flow diagram[5]

The process design review involved an analysis of both traditional and non-traditional separation methods such as distillation, liquid, liquid extraction, extractive distillation, reactive distillation, molecular sieve adsorption, and membrane pervaporation. ASPEN Plus was utilized to develop design models for the equilibrium-based separations. Conventional distillation is limited in obtaining pure IPA from the waste due to the fact that water forms an azeotrope with IPA at 87.4 wt%. The azeotrope is not pressure sensitive, which limits the use of pressure-swing distillation. Distillation to the azeotrope was proposed as the first step in sequential separations train followed with a more advanced separation operation. This effectively leverages the advantages of each technology. There is no "magic bullet" for a one step recovery operation.

Extractive distillation was evaluated for the combined waste stream. These processes involve a tertiary phase to purify the IPA composition past the azeotropic composition. Various mass separating agents (MSA) were investigated through literature review and using ASPEN Plus. Using diisopropyl ether as the MSA produced a 99% pure IPA product , but required a high operating pressure (30 atm) distillation column. A process using ethylene glycol produced a 94% purity IPA product, but needed high column utilization (4 columns). Dimethyl sulfoxide as a MSA produces 99% purity IPA product, but also has a high column utilization (4 columns). All of these proposed alternatives exceed equipment capabilities at Barceloneta and therefore were not considered for a full-scale design analysis[6, 5].

Molecular sieve adsorption was also evaluated. Even though this is not in the equipment inventory at Barceloneta, this option was considered since the process is used widely in commercial ethanol drying. These systems are not modeled in standard ASPEN Plus, therefore vendor and literature investigations were undertaken. Typically, these systems are used for low water removal in solvent drying applications and therefore it was proposed that this process follow distillation in a design sequence. To optimize the process the azeotropic overhead from the distillation column was sent to the molecular sieve in vapor form. A dual column sieve system was used to allow for IPA production and regeneration cycles. Based on the information obtained from various design literature and vendor sources, a capital cost of $500,000 (base module) with a total installed cost of $1.5 MM was estimated. IPA product with a purity of 99.5% can be obtained.

Pervaporation was considered as one of the advanced separation methods. Various configurations were examined using the equipment available at the plant. These included:

- Pervaporation units run in series or parallel
- Distillation followed by pervaporation
- Distillation followed by pervaporation followed by another distillation column

The basis for the design is a pervaporation system using 2 pressure vessels with 3 membrane modules of 35 m^2 each (total system area = 210 m^2).

The aforementioned combinations were then modeled using the combined waste streams and just separating the segregated waste of dryer distillates and centrifuge wash. Empirical models were developed for trans-membrane mass transfer using Excel and an Aspen user block with literature data [7,8]and a simplified cross-flow model. Additional design information was obtained from Sulzer. Selling the mother liquor or incineration options were evaluated in the segregated waste recovery approach. These models have limitations in their ability to predict multicomponent transport systems such as this case with the small amounts of the methanol and ethanol.

Based on scheduling of the pervaporation system, only the segregated dryer-distillates and centrifuge wash streams are treated. The distillation-pervaporation option is predicted to yield an IPA purity of 98.4% based using as a representative design basis of 1000 kg/hr feed. Higher feed rates significantly decrease IPA product purity. To purify this further with the same membrane area and feed rate, a distillation-pervaporation-distillation scheme is proposed. The use of the "polishing" distillation column increases IPA purity to 99.1%. The pervaporation system does require additional utilities such as steam (for inter-stage heaters) and electricity for vacuum system and coolant system, but these are small when compared to the overall energy saved [5,6].

A thorough design analysis was conducted on the "base case" and the various green design alternatives, distillation-pervaporation, distillation-pervaporation-distillation, and distillation-molecular sieve adsorption. For each of these a further analysis was done on the sale of the mother liquor or incineration. A summary will be shown for the distillation-pervaporation with mother liquor sold case.

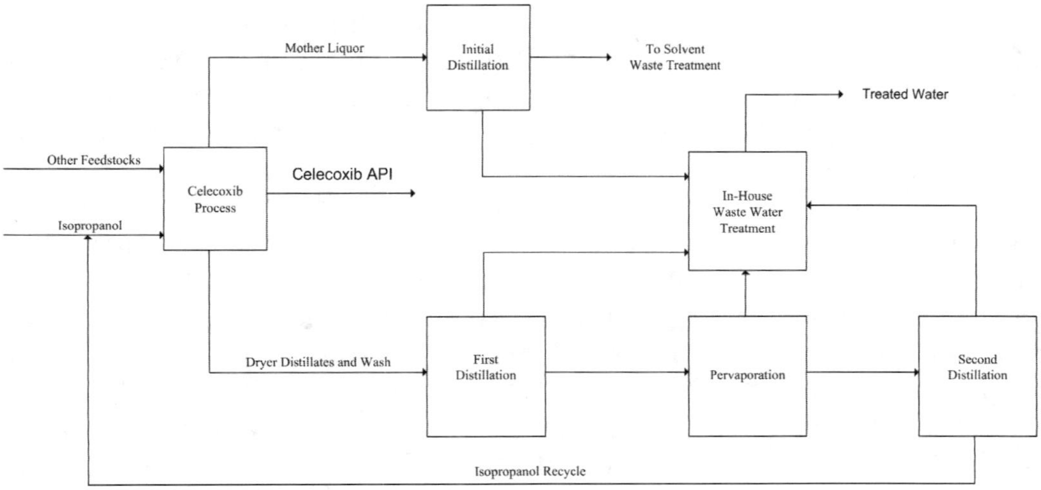

Figure 2. Proposed distillation-pervaporation-distillation process flow diagram

The process scheme shown in Figure 2, recovers 2325 kg/batch of IPA at a purity of 99.1%. This represents 57% of the IPA recovered from the total waste streams. The utilities for the proposed recovery scheme are 10,000 kg/batch of steam, 59 kWh/batch of electricity, and 91,200 gal/batch of cooling water. A comparison of annual operating costs to the base case was performed. A comparison of annual operating costs of the proposed option to the base case shows a 72% savings. The base case annual cost of $5.28 MM is reduced to $1.46 MM with the "green design".[5] The significant difference comes from the reduction in the cost to purchase fresh IPA and reduction in the waste disposal costs. There are still costs associated with the purchase of IPA, but 43% less than the base case. The cost of utilities required for the distillation-pervaporation-distillation "green design" is small when compared to the overall savings.

The economic conclusions of the green design alternatives indicate that pervaporation and molecular sieve options provide comparable operating savings (64-73%) to the base case. The molecular sieve option requires a $1.5 MM capital investment. Since the pervaporation system already is available at the plant site, this provides a non-capital intensive alternative to molecular sieve adsorption. The "best approach" in using the pervaporation option is to use the "touch-up" second distillation to get 99.1% isopropanol purity which only requires an additional 0.51% increase in operating costs, but yields a 72% savings overall. The option with selling the concentrated mother liquor decreases operating costs by 24% [5,6].

The environmental analysis of the green process designs involved several steps and only a summary of those are presented here. In order to perform a life cycle analysis (LCA), the system boundaries, purposes and functional units were chosen. This was followed by a life cycle inventory where inputs of mass and energy, and outputs of emissions and products were quantified. Finally, environmental impacts were characterized for the various "green" process alternatives.

Since isopropanol was the solvent to be recovered, the primary focus was on its life cycle. Figure 3 shows the LCA system boundaries for the proposed distillation-pervaporation-distillation and sell mother liquor option. This LCA includes the emissions produced from the manufacture of virgin IPA as well as those from the process modifications (pervaporation and distillation processes). Since mother liquor is concentrated for sale, we assumed for analysis that this could be considered generic solvent and therefore the manufacture of that would be avoided. An analysis of the proposed process to the base case indicates that emissions are reduced from 22,837 to 1,857 kg/batch. Resultant CO_2 emissions are estimated to be reduced from 20,260 to 1,066 kg/batch[5].

The overall environmental conclusions are that all pervaporation and molecular sieve adsorption options provide substantial life cycle emissions savings compared to the base case. These range from 64 to 93 % in overall emissions (and 65 to 96% CO_2 emissions) depending on the process alternative (Figure 4). Selling the mother liquor significantly reduces life cycle emissions, improving alternatives by 75%. When the recommended green process design of pervaporation-distillation-pervaporation & selling mother liquor is analyzed on a yearly basis, the following life cycle waste reductions are determined. A yearly savings of 16.63 MM kg of emissions a year, for a 92% reduction from the base case is estimated. The CO_2 reductions parallel this as well, with yearly reductions estimated at 11.55 MM kg/yr for a 95% reduction.[5]

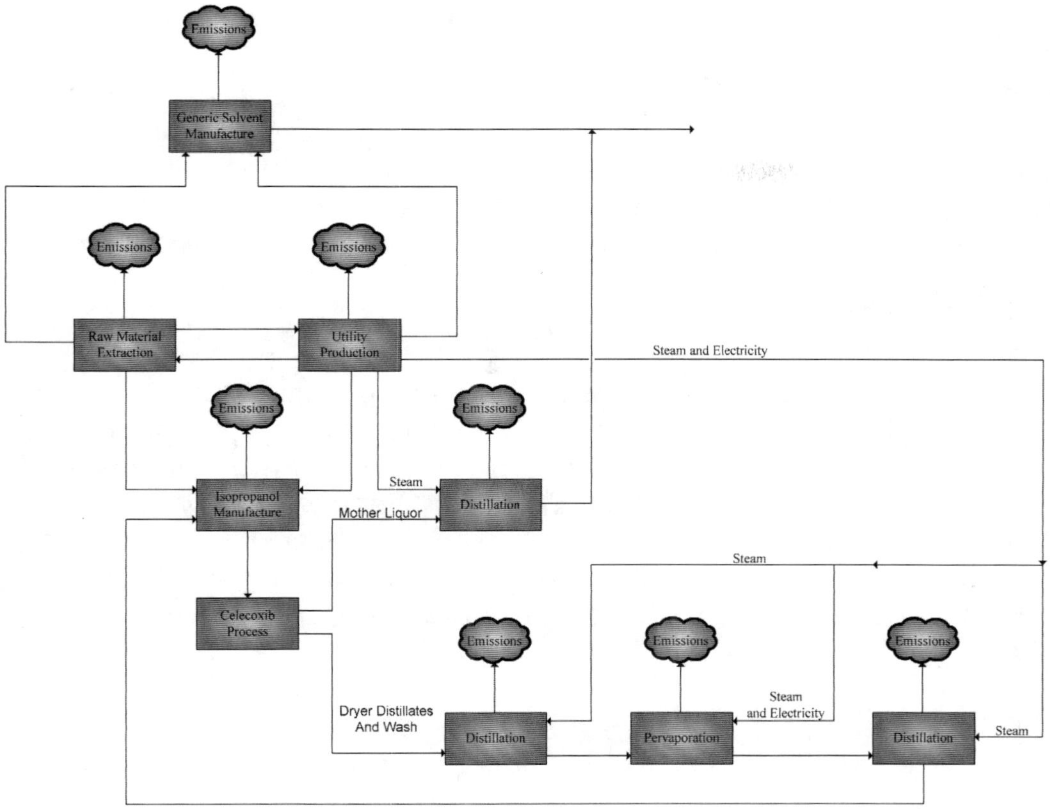

Figure 3. LCA system boundaries for the distillation-pervaporation-distillation & sell mother liquor process option.

	Dist-Incin	Dist-PV-Dist	Dist-MS	Dist-PV	Dist-PV-Dist.Sell	Dist-MS.Sell	Dist-PV.Sell
Total Emissions	34.8%	65.0%	64.4%	66.3%	91.9%	91.2%	93.1%
CO2 Emission	39.4%	66.3%	65.3%	67.5%	94.7%	93.8%	95.9%

Figure 4. Analysis of life cycle waste reductions from the various process alternatives.

CONCLUSIONS

The distillation-pervaporation-distillation process with sale of mother liquor is the recommended green engineering alternative. This option takes advantage of the current capital equipment at Barceloneta and could be implemented with minor equipment modifications such as the purchase of newer membranes. Further pilot testing is recommended to verify the models presented.

ACKNOWLEDGEMENTS

The authors acknowledge the Rowan Engineering Clinic students: Anthony Furiato, Kyle Lynch and Timothy Moroz for their tireless efforts in this project. The authors appreciate the guidance provided by Mr. Jorge Belgodere of the Pfizer Barceloneta plant. The authors thank the Pfizer Green Chemistry program and Pfizer Green Chemistry Lead, Dr. Peter Dunn for their support of the Rowan clinic partnership. The authors acknowledge the support of the U.S. Environmental Protection agency through the Pollution Prevention grant program (NP97257006-0) for their support of the Rowan projects with the green engineering in pharmaceutical manufacturing initiative.

REFERENCES

1. Anon, "Students, Pfizer improve pharmaceutical process", *Reliable Plant*, www.reliableplant.com/article.asp?articleid=12078 ,June 2008.
2. Scot, A, "Pfizer seeks to reduce solvent use," *Chemical Week*, http://www.chemweek.com/people_and_business/green_issues/12391.html, June 2008
3. Flavell-White, C, "Rowan University to cut Pfizer's solvent use," *TCE Today*, www.tcetoday.com/tcetoday/NewsDetail.aspx?nid=10530, June 2008.
4. Slater, C.S.; M. Savelski, B. Lefebvre, S. Farrell, R. Hesketh, "Using the Rowan University engineering clinic model to implement green engineering partnerships with the pharmaceutical industry," *12th Annual ACS Green Chemistry & Engineering Conference*, Washington, DC, June 2008.
5. Savelski, M.J, C.S. Slater, G. Hounsell, D. Pilipauskas, F. Urbanski, "A collaborative partnership to develop strategies for waste minimization and solvent recovery for the celecoxib process," *12th Annual ACS Green Chemistry & Engineering Conference*, Washington, DC, June 2008.
6. Furiato, A.; K. Lynch, T. Moroz, M.J. Savelski, C.S. Slater, "Green engineering design for the recovery of solvents in API manufacturing," *12th Annual ACS Green Chemistry & Engineering Conference*, Washington, DC, June 2008.
7. Qiao, X.; Chung, T.; Guo, W.F.; Matsuura, T.; Teoh, M.M., "Dehydration of isopropanol and its comparison with dehydration of butanol isomers from thermodynamic and molecular aspects". *Journal of Membrane Science* (2005), 252, 37-49..
8. Van Hoof, V.; Van den Abeele, L.; Buekenhoudt, A.; Dotremont, C.; Leysen, R. Economic comparison between azeotropic distillation and different hybrid systems combining distillation with pervaporation for the dehydration of isopropanol. *Separation and Purification Technology* (2004), 37(1), 33-49.

290c Case Study of a Greener pH Swing Crystallization In a Double Feed Stirred Reactor

Daniel Sujo[1], Robert P. Hesketh[2], and Dr. Stephanie Farrell[2]. (1) Rowan University, Glassboro, NJ 08028, (2) Chemical Engineering, Rowan University, 201 Mullica Hill Rd, Glassboro, NJ 08028

An industrial crystallization system was analyzed using a green engineering approach. The current method to crystallize a metal catalyst uses a stirred reactor, where an acid solution is added to the mother liquor to precipitate the catalyst in a semi-batch fashion. A double-feed reactor configuration was also examined and found to improve this process. This new scheme reduces the turnaround batch time by 90% and the environmental impact by 30% (LCI metrics), compared to the current single-feed operation scheme. The experiments showed that operating in a double feed mode allowed the crystallization to be conducted at a constant pH which resulted in a high purity, larger particle size. From this, a high settling velocity was achieved and resulted in a reduction in batch time. Additional experiments showed a relationship between the pH and the size of the agglomerates and the purity of the crystals. Further research on the crystallization mechanism and kinetics is strongly recommended for the generalization of the concept found in this investigation or its application as a continuous process.

290d Improving the Synthesis-to-Hydrolysis Ratio for Beta-Lactam Antibiotics Using Membrane Reactors

Andria L. Deaguero and Andreas S. Bommarius. Chemical and Biomolecular Engineering, Georgia Institute of Technology, 315 Ferst Drive, Atlanta, GA 30332

Beta-lactam antibiotics, such as ampicillin and amoxicillin, comprise 65% of the total antibiotics market with annual sales of $15 billon (Chandel:2008). Currently, ampicillin and amoxicillin are produced on large scale by condensation of D-phenylglycine ester and p-hydroxy-D-phenylglycine ester, respectively, with the beta-lactam nucleus (6-aminopenillanic acid, 6-APA) catalyzed by penicillin G acylase (PGA) (Arroyo: 2003). The condensation reaction can be carried out under thermodynamic or kinetic control, but typically kinetic control is preferred as it results in higher product yield. A potential disadvantage of the kinetically controlled process is the hydrolysis of the side chain ester and secondary hydrolysis of the semi-synthetic beta-lactam. For ampicillin, the synthesis-to-hydrolysis ratio is about 2.8 in water at 20°C and a pH value of 7.5 (Bruggink:1998). Finding a method to increase the synthesis-to-hydrolysis ratio would contribute to increasing the availability of effective antibiotics while enhancing the environmental friendliness of the process.

Our goal is to improve the synthesis-to-hydrolysis ratio of ampicillin to 5 (in case of kinetically-controlled synthesis) or the degree of conversion to ampicillin (in case of thermodynamically-controlled synthesis) with the use of a membrane reactor system. The system under investigation consists of a packed bed of immobilized PGA in series with a selective membrane that allows us to control the amount of water in the reaction mixture. The nonporous high-flux perfluoro membranes that we employ have a large free volume space, which results in high permeation rates, but are only permeable to water, hydrogen, and other small molecules. While removing water from the reaction system suppresses the hydrolysis reaction, it is necessary to maintain a certain amount of water for enzyme function. Therefore, the amount of water in the system is carefully monitored using known membrane properties and transmembrane pressure.

The presentation will focus on the condensation of phenylacetic acid and 6-APA to ampicillin. We will report the dependence of reaction rate and yield on solvent composition, especially water content, and compare our system with the conventional process.

KmX is a clean technology company providing an alternative to disposing waste chemicals through incineration. There are approximately 55,000 chemicals in use today with approximately 1,000 new compounds formed each year. The most widespread methodology for disposing spent chemical streams is through incineration and landfill. We believe that from an environmental, public health as well as economic standpoint there is a need to provide a more comprehensive plan where recycling chemical waste streams would play a more prominent role than it does today. This would allow us to develop a more sustainable waste minimization program through the recovery and rectification of spent chemicals. A comprehensive plan for minimization should include the exploration and development of products and processes that do not use toxic chemicals in the first place. However, given that we are still some way off from the ubiquitous use of green chemicals, recycling toxic and hazardous waste streams to recover the inherent value within these streams would seem a very good second best.

The "Toxic Release Inventory" (TRI 2006) data provided by the US Environmental Protection Agency (EPA), found that 24.5 billion lbs of waste chemicals were generated in 2006 from a sample of just under 600 tracked chemicals. Considering the EPA sample is just approximately 1% of the estimated total number of chemicals in use today, we start to realize the immense challenge we face annually in finding suitable disposal methods for this toxic waste.

The biggest restraints to recycling being a more prominent solution in the recovery of spent chemicals have been a lack of technology, government regulations that have worked against recycling solutions as well as insufficient data about the environmental benefits of rectification versus incineration.

The advent of pervaporation membrane technology is providing a new dimension in the pursuit of increasing the role recycling plays in the disposal of spent chemicals. To be of value the technology must be capable of providing separation capabilities that are far superior to those separation technologies that exist today as well as being totally scalable. After decades of research and development since P.A Kober invented it in 1917, we believe pervaporation membrane technology is finally at a stage where it can be put to commercial testing and should provide a paradigm shift in the way we dispose of spent chemicals. In addition certain changes taking place shortly to EPA regulations surrounding the liability of hazardous waste disposal should drive more spent streams from incineration to recycling.

In the past few years there have been several studies showing the benefits of rectification compared to incineration. One such study is the "Environmental Assessment of Waste-Solvent Treatment Options" by Christian Capello and Stefanie Hellweg. This study showed that the rectification using distillation rather than incineration of most waste solvents would benefit the environment with the key determinants being yield loss and levels of purification attained. Two studies performed in the US in 1993 and 1994 show that if material that is currently marketable and recyclable --- which is typically burned in a modern trash incinerator --- was recycled instead, some 3-5 times as much energy would be saved. The reason: Incineration can only recover some of the calorific value contained in the waste stream; it cannot recover any of the energy invested in extraction, processing, fabrication and chemical synthesis involved in the manufacture of the objects and materials in the waste stream.

We will review how hybrid distillation/pervaporation membrane-aided solutions will allow for a much broader spectrum and volume of waste chemicals to be recovered than hitherto possible thereby allowing for a far greater role to be played by recycling in the disposal of waste chemicals.

290f Thermally and Chemically Resistant Novel Perfluoromembrane Reactors for Enhancing Reaction Rates and Conversion for Equilibrium-Limited Reactions

Stuart Nemser[1], Sudip Majumdar[1], Joseph Fox[2], Maksim Royzen[2], and Andrew DeAngelis[2]. (1) Compact Membrane Systems, Inc, 335 Water Street, Wilmington, DE 19804, (2) Chemistry & Biochemistry, University of Delaware, Newark, DE 19716

For equilibrium-limited reactions, selective removal of a by-product can create an environment in which significantly higher conversions of the reactants are possible. Many syntheses involve condensation or dehydration reactions that generate water as a by-product. This water must be removed since (1) water is often an undesirable component of the finished product, (2) water limits conversion to final product, (3) water generated can slow the reaction rate, and (4) presence of by-product water enhances undesired side reactions. Many syntheses reactions have thermal limitations and therefore removing the water by boiling it off is simply not attractive due to product and/or catalyst degradation. Membrane reactors have been proposed for a number of chemical syntheses. To be competitive with conventional technologies, membrane reactors must be shown to have better selectivity, permeability, stability and ability to perform over a wide temperature range. Compact Membrane Systems (CMS) has identified a novel product concept that represents a broad platform for utilization of membrane reactors in chemical synthesis. Basic data in combination with preliminary economic and engineering analysis show that removal of water will significantly increase yield and therefore lower reagent consumption. Further, the desired product downstream separation and purification can be dramatically simplified. Therefore, enhanced water removal reduces production costs up to three ways: higher yield, faster reaction and purer product. CMS will present membrane reactor performance and separation data over a broad temperature and chemical composition range.

Highly permeable, chemically and thermally resistant perfluorinated membranes were used to selectively remove water from reaction mixtures. The process operated by contacting the reaction mixture with a perfluoropolymer membrane. A vacuum was applied on the other side of the membrane to provide a driving force for the water transport. Product conversion beyond the conventional equilibrium limitation was achieved. Significant enhancement in reaction rate was also observed. The impact of applying this novel membrane-based technology on the reaction product will be discussed.

351a Green Engineering Activities In the Pharmaceutical Industry

Nhan T. Nguyen, Office of Pollution Prevention and Toxics, US EPA, 1200 Pennsylvania Ave., NW, Washington, DC 20460 and Sharon Austin, Oppt, US EPA, 1200 Pennsylvania Ave., NW, Washington, DC 20460.

This presentation will provide an overview of the Green Engineering Program within the Office of Pollution Prevention and Toxics of the US EPA, collaborative activities with other partners in academia, the regions, industry and governmental agencies to reduce the environmental footprint of the pharmaceutical industry. The presentation will also discuss potential opportunities within the pharmaceutical industry based on recent activities primarily addressing the impact of waste solvents.

Web Page: www.epa.gov/oppt/greenengineering

351b Green Chemistry and Sustainable Futures Activities within the Pharmaceutical Industry

Richard E. Engler, Green Chemistry Program, US Environmental Protection Agency, 1200 Pennsylvania Ave. NW (7406M), Washington, DC 20460, Kelly Mayo, Epa/oppt, US Environmental Protection Agency, 1200 Pennsylvania Ave. NW (7406M), Washington, DC 20460, and David J. C. Constable, Corporate Environment, Health & Safety, GlaxoSmithKline, One Franklin Plaza, P.O. Box 7929, Philadelphia, PA 19101-7929.

The presenter(s) will provide overview of OPPT's Green Chemistry and Sustainable Future Programs. Ideal sets of solvents from a life cycle and environmental perspective will be reviewed. Some selected tools which may be useful for estimating physical chemical properties and selection of alternative solvents will be covered.

Green Pharma

Do the right thing ... *and* save money!

WHY SHOULD MANUFACTURERS ACTIVELY MANAGE ENVIRONMENTAL IMPACTS?

Environmental performance contributes to the overall success of any organization. Managing the environmental aspects of your business — such as energy and materials use — will increase your profitability by offering:

- **Cost Savings & Operating Efficiency**
  Reducing waste and conserving energy and water can cut operating costs, improving the bottom-line of your organization.
- **Increase Market Share in the Global Arena**
  Progressive environmental programs position your company to meet international standards in the global supply chain.
- **Increase Shareholder Value**

  Protect your brand by reducing your liability and risk exposure by using less toxic alternatives.
- **Succeed in Hiring and Retention of Talent**

  Recent surveys show MBAs and engineers rate corporate social responsibility highly in selecting positions.

GREENING PHARMACEUTICAL MANUFACTURING

Due to the large energy, water and materials use in pharmaceutical manufacturing, efficiency is a key strategy for profitability. Taken together, plants and office buildings share some of the same environmental concerns as a city. The Environmental Protection Agency's (EPA) Green Pharma initiative helps facilities in improving environmental performance in specific areas. The Green Chemistry in Pharmaceutical Manufacturing initiative was put together with key industry leaders, research universities and representatives from the Food and Drug Administration.

GREEN PHARMA OVERVIEW

EPA provides facility managers with clearly defined opportunities for becoming a recognized leader in green operations. The Green Pharma initiative provides a portal to many types of voluntary technical assistance and is based on a relationship of trust between the EPA and the facility. It offers a way to join the growing community of conscientious corporate citizens and improve stakeholder value. Facility managers can gauge performance, set goals for improvement and get recognition for their results.

TECHNICAL ASSISTANCE AVAILABLE

Upon request, EPA can assist with identifying pollution prevention opportunities specific to your facility, provide references for case studies in your region, and locate additional contacts and resources within your community. EPA offers non-regulatory assistance in areas such as:

- Energy Efficiency
- Resource Conservation
- Green Chemistry
- Water Conservation
- Compliance Assistance
- Green Supply Chain
- Green Buildings
- Carbon Reductions

☐ *Yes!* **I WOULD LIKE MORE INFORMATION.**

Join the dynamic community of business leaders, professional societies, academia and green service industries to identify and implement the best environmental practices for your operations.

Become a part of EPA's Green Pharma efforts by contacting Maryann Helferty at 215-814-3299.

GREEN PHARMA OPPORTUNITIES

Being a Green Pharma partner provides facility managers with clearly defined pathways to measure environmental performance and set goals for improvement. Your efforts in these performance areas determine your level of success:

PERFORMANCE AREAS	GREEN PHARMA OPPORTUNITIES
Energy Efficiency	▪ Plants can make energy management plans, upgrade to efficient equipment, and choose effective facility improvements. Optimize process heating and cooling systems. Take a new look at cogeneration opportunities.
Water Conservation	▪ Facility implements water conservation practices in operating systems, process design and landscaping practices. Join EPA's Water Sense or Greenscapes programs.
Renewable Energy	▪ Facility is an EPA Green Power Partner or purchases power from a renewable source, such as solar energy, geothermal, biomass, or wind.
Reduced Use of Toxics	▪ Facility eliminates or reduces toxic chemicals from products, waste streams, and industrial releases. Get help with pollution prevention, and waste minimization. Join EPA's National Partnership for Environmental Priorities (NPEP).
Solid Waste	▪ Facility sets and achieves a waste reduction goal, encouraging re-use and recycling. Join EPA's Waste Wise.
Green Supply Chain	▪ Operation has an established green procurement policy.
Climate Change	▪ Learn about EPA Climate Leaders and get assistance in doing a greenhouse gas inventory and making a reduction plan.
Environmental Messages	▪ Facility develops and implements an outreach program to communicate environmental programs to customers and staff.
Green Buildings	▪ Identify and implement high performance standards for new and existing buildings for higher efficiency and savings.
Professional Recognition	▪ Join an EPA program, such as Waste Wise, Performance Track, Green Power, Climate Leaders, NPEP, or Water Sense.
Global Certification	▪ Develop an Environmental Management System to meet new international standards for international marketing and sales.
Community Education	▪ Improve business climate with regional greening initiatives for quality of life such as Green Communities and SmartWays.

A Green Pharma information packet that includes more details about each performance area and assistance is available by contacting Maryann Helferty at 215-814-3299.

Interagency Network of Enterprise Assistance Providers- An unusual partnership of government agencies and private businesses communicate and collaborate to enhance services to small business.

In today's interconnected world collaboration needs to be routine and continuous. In 2006, this need led to the formation of a more formal public/private network, the Interagency Network of Enterprise Assistance Providers (INEAP).

INEAP is a coalition of people from federal (and some private and nonprofit) organizations whose primary mission is to assist small businesses. It has three major goals:

1. Establish and maintain positive relationships among the people who run these programs.
2. Use those relationships to learn about the various programs—what they offer, whom they serve, and what if any collaborative relationships they currently have with related programs.
3. Use that knowledge to find opportunities to collaborate and improve services to small businesses.

INEAP members come from over forty-five business and technical assistance programs, including the SBA; Export- Import Bank; U.S. Departments of Commerce (DOC),Labor (DOL), and Energy (DOE); and U.S. Environmental Protection Agency (EPA). The U.S. Women's Chamber of Commerce, American Association of Community Colleges (AACC), and other nonprofit organizations are also involved.

INEAP's stakeholders are other federal, state, and private providers that offer services to or have a network with small and medium-size businesses. Congress is also a customer and stakeholder.

Fifteen to forty INEAP members, depending on the agenda, meet monthly. The meetings typically help members learn about other programs' offerings. Each meeting focuses on a theme: support to veteran-owned businesses, international trade and financing, loan programs to businesses, and the like. Between meetings, various INEAP members forge the formal and informal partnerships that actually accomplish the network's goals.

INEAP has no senior sponsor, no formal charter, and no official mandate. It operates with a few ground rules (such as respect members and each other's programs and commit to sharing program information), but compared with some partnerships, it functions in a very informal manner. Members prefer it that way because the low profile keeps bureaucracy to a minimum.

Antonio Doss, the director of SBA's Small Business Development Centers, and Carroll Thomas, partnership catalyst for MEP and author of this article, founded this interagency group and facilitate the monthly meetings. We recently established an executive core council of the most active INEAP members, which shares the leadership tasks. Meetings rotate among the members' downtown offices. When resources (people, meeting rooms, and the like) are needed, the leaders ask members to contribute them. High technology has not played a major role in the network's operation, but INEAP does make occasional use of the Web, and its Web site— www.ineap.nist.gov/ineap/home.ineap —has an external and internal component. The internal area is for members to share information—on topics such as business disaster preparedness, computer recycling programs, and energy efficiency—with other members and their clients. INEAP also plans to offer a series of Webcasts on assistance programs for the taxpaying business owners its organizations serve. The major communication tool INEAP members use is an old-fashioned one—meeting and talking directly with each other to exchange information and find common purpose.

On September 26, 2007, fifteen INEAP members and visitors gathered at DOE in Washington, DC, for the monthly meeting, one that had a particular sense of energy and enthusiasm. Participants were from DOL, DOE, DOC, four EPA program offices, and the AACC. Members' presentations took up much of the time, but this meeting differed in the connection and synergy apparent in the room. One member discussed a partnership among several agencies and private-sector groups to reduce energy usage by 25 percent over the next decade, and its plan to post its methods, tools, and progress on a Web to share them with other companies. That got the attention of an EPA representative, who wanted to know more about the partnership and how the members were working together. That member then mentioned a consortium of college presidents who are meeting to find ways to reduce energy consumption on their campuses, which prompted someone else to talk about an upcoming conference on best practices in energy usage and lean methods within government agencies.

Green Suppliers Network

Before INEAP, MEP and EPA established the Green Suppliers Network (GSN), a collaboration between government and industry that focuses on the dual challenge of reducing the environmental impact of small and midsize manufacturing suppliers and increasing the companies' efficiency, productivity, and profitability. GSN employs "lean" and "clean" methods, which concentrate on the root causes of waste in a facility and provide a framework for achieving specific measurable environmental and business objectives. GSN helps companies in three areas: lean processes, reduced energy consumption, and reduced impact on the environment.

In Missouri, MEP had been working with EPA to identify companies interested in GSN reviews. At the same time, it was working with the local DOE industrial assessment center to identify and assist companies interested in energy assessments. And it was providing lean assessments to identify waste in processes and materials. The problem was that all three of these activities were happening separately. Jimmy Story, business program manager of the Missouri MEP, was looking for ways to offer the GSN package of three assessments as one package that made sense to the manufacturers. At an INEAP meeting in Washington, MEP and EPA program managers discussed the need to bring the three GSN elements together and learned that Jimmy Story was looking for the same thing at the field level. Through INEAP connections, Jimmy Story was able to get encouragement from several government program managers. He also received additional assessments from DOE (which provides them at no charge to companies) and learned how to seek additional funding from EPA to help defray the cost of environmental assessments.

The biggest winners of all are the fifteen manufacturers that are current Missouri MEP clients. Getting lean, clean, and energy efficient helped them to become more profitable and save on the cost of the assessments. Having the assessments rolled into one also cut the downtime of going through each assessment separately. Most important, integrating the assessments helped them save energy and improve the air quality of the plant. For more on GSN, go to www.greensuppliers.gov/gsn/home.gsn.

Supporting and leading the partnership takes an enormous amount of time, due, in part, to the large number of established relationships. The leaders know investing time is key in relationship-building efforts and communicating these relationships. This groundwork allows the members to contact someone with whom they have an established relationship, cutting through the normal red tape of government-structured stovepipes.

In addition, each meeting requires preparation to pick appropriate themes and involves time-consuming follow-up afterward. "All INEAP members do this work on their own time," Doss notes. "Nobody's boss directs the members to participate. "That's testimony to the partnership's attractiveness. It's also a risk: if the members' managers have no stake in INEAP, they may not approve of the amount of time their subordinates put into the partnership. Happily, every INEAP member's supervisor has seen the value of this partnership and has supported the investment of time.

Two people perform much of the INEAP work. "Carroll is really the glue that keeps everything together," Doss says. "I do a lot, but she's critical to our success. So we need to develop an infrastructure to keep things moving if she (or I) can't continue contributing at this level." The recently created council will not only expand the number of people who feel ownership of the group, but will encourage succession planning for developing a new generation of leaders to assume some of the workload and continue efforts into the future.

Excerpted from "The Public Manager" spring 2008 Russ Linden and Carroll Thomas

Authors: Carlos Ramos, Team Leader, Pollution Prevention Team, U.S. EPA Region 2, Eduardo González, Chemical Engineer, Caribbean Environmental Protection Division, EPA Region 2, and Walter Schoepf, Environmental Scientist, Pollution Prevention Team, U.S. EPA Region 2, New York, New York.

Title: Promoting Environmental Stewardship through Industry /Government Collaborative Partnerships

Abstract: The phamaceutical industry has the highest waste generation per pound of product produced and the highest amount of organic solvents used per pound of product produced for any commercial sector. The majority of drug products are made through organic synthesis processes that require many sequential "batch" reaction steps, that require large quantities and number of organic solvents, which can pose significant environmental concerns regarding release into the environment. These solvents have varying degrees of toxicity to humans and animals, and some solvents are known or possible carcinogens. With that in mind, and for the reason that the pharmaceutical industry is a very important "cluster" in our regional economy, including Puerto Rico, Region 2 has been collaborating with Region 3, the Office of Pollution Prevention and Toxics, and the Food and Drug Administration, and other industry stakeholders to promote pollution prevention and stewardship in the pharmaceutical sector through the implementation of green chemistry and green engineering concepts and tools. The collaborative effort will require 4 major components: partnership building, outreach and education, direct technical assistance, and in certain circumstances financial support to other organizations that are leading "change agents" for greening industry practices. The section will describe: activities sponsored to date, the regional role within the collaborative partnership, and emerging efforts and perspectives related to building a results oriented government - industry collaborative partnership in Puerto Rico. The goal of this Puerto Rico pilot project is to build upon and enhance, as well as support ongoing efforts by industry to reduce, recover, reuse and catalyze third party market exchanges of solvents. Solvents must be considered as a secondary product, not waste. EPA R2 hopes that the pilot effort in Puerto Rico will highlight the: (a) business case for the reduction, recovery and reuse of these solvents in chemicals, energy, emissions and other economic and environmental benefits, (b) and clarify and/or remove any barriers, perceived or real, on the reduction, recovery and reuse of these solvents as necessary.

351f Overview of FDA's Process Analytical Technology (PAT) Process and Green Product Quality

Christopher Watts, FDA, Office of Pharmaceutical Science, CDER, Room 3548, Silverspring, MD 20993

This presentation will cover FDA agency wide initiative of Process Analytical Technology (PAT), and opportunities for "Green" quality and control within pharmaceutical processes that are registered with the FDA. Discussion will also cover the ability to perform process optimizations to registered products and processes, and utilizing process optimization to infer product quality by minimizing variations within the process.

TREATMENT OF ENDOCRINE DISRUPTING COMPOUNDS IN WATER

Shammi Rahman, Justin Moman, Dr. Robert A. Angus, and Dr. Robert W. Peters
University of Alabama at Birmingham, Birmingham, AL

Abstract

Historically, compounds such as substituted phenols, non-biodegradable chlorinated solvents, pesticides and surfactants, are recognized as examples of substances that are difficult to remove from water. Recently, pharmaceuticals and personal care products (PPCPs) and especially endocrine disrupting compounds (EDCs) are considered as emerging contaminants, which means that they are still unregulated or in the process of formulating regulations.

There is growing evidence of the impact of these emerging contaminants in the environment. Studies have shown that male fish in detergent-contaminated water express female characteristics, turtles are sex-reversed by polychlorinated biphenyls (PCBs), male frogs exposed to a common herbicide form multiple ovaries, pseudohermaphroditic offspring are produced by polar bears, and seals in contaminated water have an excess of uterine fibroids. Recent work shows that human development can also be feminized by exposure to estrogenic chemicals.

As an example, passive sampling is highly complimentary to spot sampling in environmental analysis. A polar organic chemical integrative sampler (POCIS) was extensively tested to optimize its performance under both controlled and field conditions. Under laboratory conditions, the kinetics of compound uptake by POCIS were linear during 10-day of exposure. POCIS sampling rates of the target compounds were significantly greater using polyethersulfone instead of polysulfone membrane, and was enhanced with increasing sorbent exposure area. Both spot and passive sampling demonstrated that most of the target chemicals were frequently detected in sewage effluent and river waters, and that the daily changes in the pollutant concentrations were greater for pharmaceuticals than for EDCs. The aqueous concentrations of all compounds were elevated at a sewage outfall, which has been confirmed to be an important source of the target compounds in rivers. The validated POCIS was successfully used to estimate the concentrations of the target compounds in effluent and river water, which were in good agreement with those from spot sampling for pharmaceuticals.

Advanced oxidation processes (AOPs) constitute a promising technology for the treatment of wastewaters containing pharmaceuticals and personal care products (PPCPs) and especially EDC's. AOPs are characterized by the generation of hydroxyl radicals. Besides fluorine, the hydroxyl radical is the strongest known oxidant. Therefore, it is possible for the hydroxyl radical to oxidize and mineralize almost every organic molecule into carbon dioxide (CO_2) and inorganic ions. The most common advanced oxidation technique used at the industrial level is the Fenton's reagent process, which uses an iron salt and hydrogen peroxide.

The challenge of EDC's is that some exist in a steroid ring that is very difficult to degrade by standard methods used in waste water treatment systems to remove pathogens. 17α-ethinylestradiol (EE2), the synthetic estrogen commonly found in birth control pills, is one such compound. Using AOP's (Fenton's reagent, hydrogen peroxide and ozone) in combination with other prospective treatment methods (sonication and UV), a solution of EE2 was subjected to treatments of varying methods, durations and intensities. These treated solutions were analyzed using high performance liquid chromatography (HPLC) in order to evaluate their effectiveness in degrading the endocrine disrupting compound EE2 in wastewater.

Introduction

In recent years, it has been determined that various synthetic and natural compounds can mimic or interfere with the action of natural hormones and disrupt the endocrine systems of humans and wildlife [Colburn *et al.*, 1996]. These substances, collectively referred to as endocrine-disrupting compounds (EDCs), have been linked to a variety of adverse effects in both humans and wildlife [Tyler *et al.*, 1998; Jobling *et al.*, 1998; McLachlan and Arnold, 1996; Sumpter, 2005]. Numerous EDCs, most of which act as estrogens, have been detected in various surface waters and ground waters [Kolpin *et al.*, 2002].

A major source of estrogens in rivers is treated wastewater effluent. Investigations worldwide have detected bioactive estrogens in waters receiving treated wastewater [Snyder *et al.*, 2001; Murk *et al.*, 2002; Sheehan *et al.*, 2002; Tilton *et al.*, 2002; Gomes *et al.*, 2003; Kolodziej *et al.*, 2003; Cargouet *et al.*, 2004; Hemming *et al.*, 2004; Sarmah *et al.*, 2006; Lishman *et al.*, 2006]. Municipal wastewater is a complex mixture of natural and synthetic organic chemicals. The most powerful EDCs commonly detected in treated wastewater include the natural hormones 17β-estradiol (E2) and estrone (E1), and 17α-ethinylestradiol (EE2), a synthetic estrogen used in birth control pills. Other non-steroidal organic chemicals have been shown to possess estrogenic activity, but are much weaker than the steroid hormones. These include the degradation products of nonionic surfactants, such as alkylphenol-polyethoxylates, and plasti-cizers, such as bisphenol A [Barber *et al.*, 2000].

Current technologies for treating municipal wastewater are only partially successful at removing EDCs [Gomes *et al.*, 2003; Svenson *et al.*, 2003; Clara *et al.*, 2005; Falconer *et al.*, 2006; Escher *et al.*, 2006].. Steroid hormones are especially difficult to remove completely. They contain steroid rings that are resistant to degradation by the microorganisms used in wastewater treatment plants [Johnson and Sumpter, 2001]. Not only are steroids not efficiently removed from the waste during treatment but, those not removed are actually activated by the treatment. Steroids are excreted in human waste as biologically inactive glucuronide or sulfate conjugates, but are hydrolyzed back to the active native molecule by microbial activity in the treatment plant [Johnson and Sumpter, 2001]. As a result, both natural and synthetic estrogens and their degradation products tend to pass through wastewater treatment systems in bioactive forms and can reach concen-trations in receiving waters sufficient to produce deleterious biological effects on organisms living in the waters [Metcalf *et al.*, 2001; Jobling *et al.*, 2002]..

Reports of EDCs in water have raised substantial concern among regulatory agencies as well as operators of wastewater treatment facilities and water purification plants. It is clear that current methods often fail to remove steroids efficiently. It would be of great benefit to the health and well being of aquatic organ-isms, as well as humans living downstream from wastewater treatment plants (WWTPs), if practical methods could be developed to increase the efficiency of removal of EDCs during the treatment of municipal wastewater. Treatment technologies, such as activated carbon and reverse osmosis appear to be capable of removal of many trace contaminants. However, they are costly to install and maintain. Research is needed to better understand the fate of these compounds during wastewater treatment, removal kinetics, and to develop less expensive treatment alternatives.

The research had two major objectives: (1) to investigate the effectiveness of various advanced oxidation process (AOP) techniques (used singly or in combination) in degrading the potent steroid EE2 in aqueous solution, and (2) to analyze water and sediments in rivers in the Birmingham area receiving treated wastewater to determine whether the estrogens are present in concentrations sufficient to disrupt endocrine systems of aquatic organisms in the receiving waters.

Methods

Advanced Oxidation Processes

Experiments were conducted using 1.0-L volumes of deionized water spiked with 1.0 µM 17α-ethinylestradiol (EE2, Steraloids, Inc., Newport, RI, USA). After treatment, 100 mL of the water was passed through a Varian Speck C18 10 mL column (35 mg sorbent). The columns were eluted with 5 mL HPLC grade methanol, dried under nitrogen gas, and reconstituted to a volume of 100 µL with HPLC grade methanol and subjected to HPLC analysis as described below.

Treatments were done for 5, 10, 15, and 20 minutes and consisted of: Sonication (1.4 kHz), H_2O_2 (1,000 ppm), UV (16 lamps, 5 W each), O_3 (100 ppm) and the following combinations: Sonication + H_2O_2, Sonication + O_3, Sonication + UV, UV + O_3, UV + H_2O_2, UV + O_3 + H_2O_2, and Sonication + UV + O_3 + H_2O_2.

Analysis of River Water Samples

Samples were taken from local rivers (Cahaba River and Buck Creek) up- and downstream from wastewater treatment plants. Samples were collected in 2 L acid-washed glass jars with 15 mL of methanol added. They were immediately placed on ice and transported to the lab. In the lab, 1.0 L of each sample was passed through a series of filters (Whatman 52 [7 µm], followed by Whatman GF/A [1.6 µm], followed by Millipore RW0304700 [0.45 µm]) to remove particulates. The filtered water samples were then each passed through a Varian Bond Elut C18 solid phase extraction column (500 mg sorbent). The columns were eluted with 5 mL HPLC grade methanol, dried under nitrogen gas, and reconstituted to a volume of 100 µL with HPLC grade methanol and subjected to HPLC analysis as described below.

HPLC Analysis

For HPLC analysis, 10.0 µL of the reconstituted sample was injected into a Perkin-Elmer HPLC system. The column used was a Varian CP30705 Microsorb 100 C18, particle size 3 µm, 100 × 46 mm. Organic constituents of the sample were separated using a water:acetonitrile solvent gradient. For spiked samples, the gradient started with a 1:1 water:acetonitrile mixture. The proportion of acetonitrile increased up to 70% in a linear fashion for a period of 10 min and then was increased to 100% acetonitrile and held for a further 10 min. For environmental samples, the gradient started with a 1:1 mixture, increased to 80% acetonitrile over 20 min and then increased to 100% acetonitrile for another 10 min. For both gradients, EE2 elutes from the column at about 6 min. Organic compounds eluting from the column were detected by UV absorbance at 225 nm.

Quantification of the amount of EE2 eluting from the HPLC column was done by comparison of the peak volume with that of a known sample containing 10.0 nmoles in 10.0 µL (1 mM solution). By passing a 1mM EE2 standard solution through a C18 column, extracting and reconstituting, as was done with all samples, it was determined that the process has a 78.5% extraction efficiency.

The rapidity by which the treatments broke down EE2 was quantified by estimating the first order decay equation rate constant.

Results

Advanced Oxidation Processes

Preliminary results (see Figure 1) indicate that the AOPs, performed individually, break down up to ~16% of the EE2 with 20 minutes of treatment. Combinations of treatments appear to be much more effective. Sonication+ozone+peroxide+UV+Fenton's reagent, in combination, broke down >28% of the EE2 in 20 minutes.

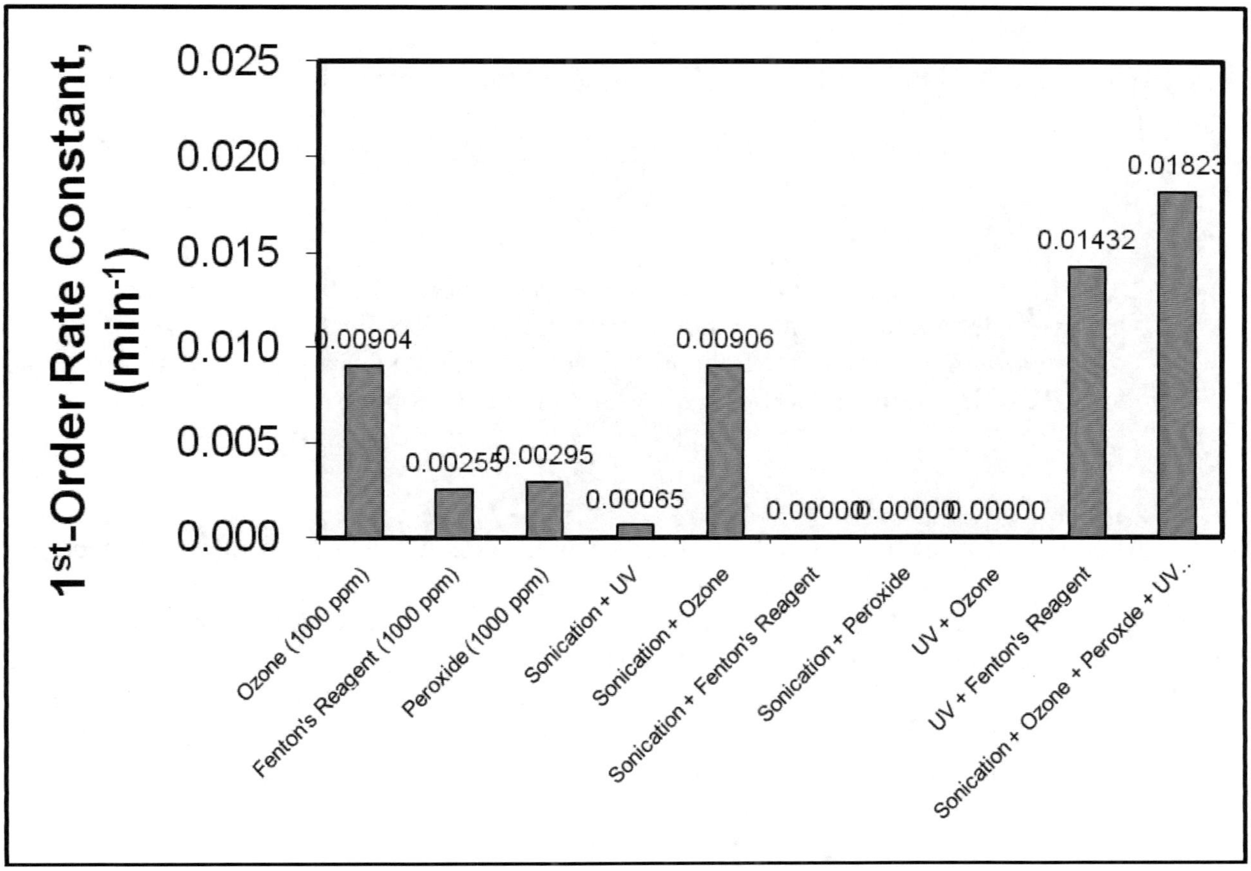

Figure 1. Estimated first order decay rate constants for the breakdown of 17α-ethinylestradiol by different advanced oxidation treatments.

Analysis of River Water Samples

Currently, samples have been collected from the Cahaba River and Buck Creek. Some of them appear, based on retention times, to have steroids in concentrations sufficient to affect aquatic organisms living in the rivers (see Figure 2). Further analysis with mass spectrometry will be necessary to confirm the identities of the suspected steroids.

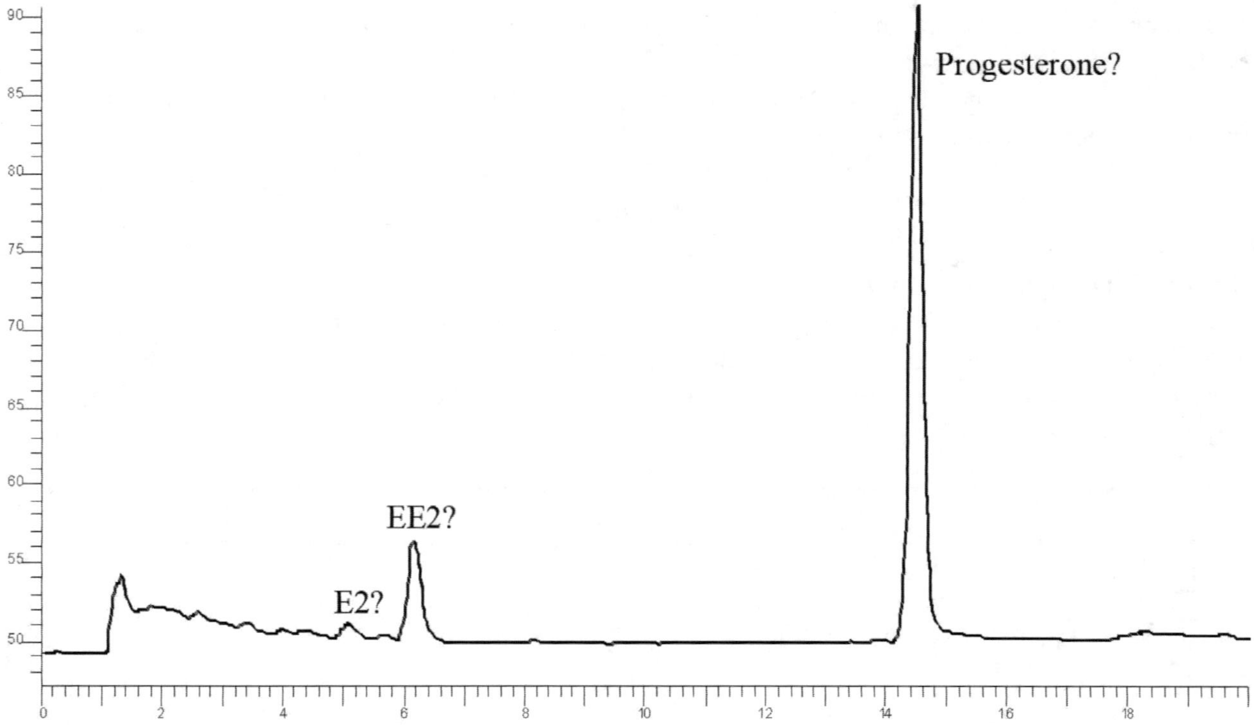

Figure 2. HPLC chromatogram of an extract of water collected from the Cahaba River below a wastewater treatment plant in Hoover. Preliminary identifications are based on similarity of retention time to known standards separated using the same solvent gradient. Definitive identification will require analysis by mass spectroscopy.

Conclusions

Although the project period is over, the project is not complete by any means. Numerous different AOP treatment combinations have been performed and await HPLC analysis. These will be done this semester by a graduate student in biology. Further river water samples will be collected and analyzed this fall as well. Preliminary results (see Figure 1) indicate that the AOPs, performed individually, break down up to ~16% of the EE2 with 20 minutes of treatment. Combinations of treatments appear to be much more effective. Sonication+ozone+peroxide+UV +Fenton's reagent, in combination, broke down >28% of the EE2 in 20 minutes.

Acknowledgments

Funding for this project was provided by the Alabama Water Resources Research Institute (AWRRI) through the U.S. Geological Survey.

References Cited

Barber, L.B., G.K. Brown, and S.D. Zaugg, 2000. "Potential Endocrine Disrupting Organic Chemicals in Treated Municipal Wastewater and River Water", pages 97–123 in ***Analysis of Environmental Endocrine Disruptors***, L.H. Kieth, T.L. Jones, and L.L. Needham (Eds.), American Chemical Society, Washington, D.C.

Cargouet, M., D. Perdiz, A. Mouatassim-Souali, S. Tamisier-Karolak, and Y. Levi, 2004. "Assessment of River Contamination by Estrogenic Compounds in Paris Area (France)", ***Science of the Total Environment*, *324***(1–3): 55–66.

Clara, M., B. Strenn, O. Gans, E. Martinez, N. Kreuzinger, and H. Kroiss, 2005. "Removal of Selected Pharmaceuticals, Fragrances and Endocrine Disrupting Compounds in a Membrane Bioreactor and Conventional Wastewater Treatment Plants", ***Water Research*, *39***(19): 4797–4807.

Colborn, T., D. Dumanoski, and J.P. Myers, 1996. ***Our Stolen Future: Are We Threatening Our Fertility, Intelligence and Survival – A Scientific Detective Story***, Dutton Books, New York, NY.

Escher, B.I., W. Pronk, M.J.F. Suter, and M. Maurer, 2006. "Monitoring the Removal Efficiency of Pharmaceuticals and Hormones in Different Treatment Processes of Source-Separated Urine with Bioassays", ***Environmental Science & Technology*, *40***(16): 5095–5101.

Falconer, I.R., H.F. Chapman, M.R. Moore, and G. Ranmuthugala, 2006. "Endocrine-Disrupting Compounds: A Review of Their Challenge to Sustainable and Safe Water Supply and Water Reuse", ***Environmental Toxicology*, *21***(2): 181–191.

Gomes, R.L., M.D. Scrimshaw, and J.N. Lester, 2003. "Determination of Endocrine Disrupters in Sewage Treatment and Receiving Waters", ***Trends in Analytical Chemistry*, *22***: 697–707.

Hemming, J.M., H.J. Allen, K.A. Thuesen, P.K. Turner, W.T. Waller, J.M. Lazorchak, D. Lattier, M. Chow, N. Denslow, and B. Venables, 2004. "Temporal and Spatial Variability in the Estrogenicity of a Municipal Wastewater Effluent", ***Ecotoxicology and Environmental Safety*, *57***(3): 303–310.

Jobling, S., N. Beresford, M. Nolan, T. Rodgers-Gray, G.C. Brighty, J.P. Sumpter, and C.R. Tyler, 2002. "Altered Sexual Maturation and Gamete Production in Wild Roach (*Rutilus rutilus*) Living in Rivers that Receive Treated Sewage Effluents", ***Biology of Reproduction*, *66***(2): 272–81.

Jobling, S., M. Nolan, C.R. Tyler, G. Brighty, and J.P. Sumpter, 1998. "Widespread Sexual Disruption in Wild Fish", ***Environmental Science & Technology*, *32***(17): 2498–2506.

Johnson, A.C., and J.P. Sumpter, 2001. "Removal of Endocrine-Disrupting Chemicals in Activated Sludge Treatment Works", ***Environmental Science and Technology*, *35***(24): 4697–4703.

Kolodziej, E.P., J.L. Gray, and D.L. Sedlak, 2003. "Quantification of Steroid Hormones with Pheromonal Properties in Municipal Wastewater Effluent", ***Environmental Toxicology and Chemistry*, *22***(11): 2622–2629.

Kolpin, D.W., E.T. Furlong, M.T. Meyer, E.M. Thurman, S.D. Zaugg, L.B. Barber, and H.T. Buxton, 2002. "Pharmaceuticals, Hormones, and Other Organic Wastewater Contaminants in U.S. Streams, 1999-2000: A National Reconnaissance", ***Environmental Science & Technology*, *36***(6): 1202–1211.

Lishman, L., S.A. Smyth, K. Sarafin, S. Kleywegt, J. Toito, T. Peart, B. Lee, M. Servos, M. Beland, and P. Seto, 2006. "Occurrence and Reductions of Pharmaceuticals and Personal Care Products and Estrogens by Municipal Wastewater Treatment Plants in Ontario, Canada", ***Science of the Total Environment*, *367***(2–3): 544–558.

McLachlan, J.A., and S.F.Arnold, 1996. "Environmental Estrogens", ***American Scientist*, *84***(5): 452-461.

Metcalfe, C.D., T.L. Metcalfe, Y. Kiparissis, B.G. Koenig, C. Khan, R.J. Hughes, T.R. Croley, R.E. March, and T. Potter, 2001. "Estrogenic Potency of Chemicals Detected in Sewage

Treatment Plant Effluents as Determined by in vivo Assays with Japanese medaka (*Oryzias latipes*)", ***Environmental Toxicology and Chemistry***, ***20***(2): 297–308.
Murk, A.J., J. Legler, M.M.H. Lipzig, J.H.N. Meerman, A.C. Belfroid, A. Spenkelink, B. van der Burg, G.B.J. Rijs, and D. Vethaak, 2002. "Detection of Estrogenic Potency in Wastewater and Surface Water with Three in vitro Bioassays", ***Environmental Toxicology and Chemistry***, ***21***(1): 16-23.
Sarmah, A.K., G.L. Northcott, F.D.L. Leusch, and L.A. Tremblay, 2006. "A Survey of Endocrine Disrupting Chemicals (EDCs) in Municipal Sewage and Animal Waste Effluents in the Waikato Region of New Zealand", ***Science of the Total Environment***, ***355***(1–3): 135-144.
Sheahan, D.A., G.C. Brighty, M. Daniel, S.J. Kirby, M.R. Hurst, J. Kennedy, S. Morris, E.J. Routledge, J.P. Sumpter, and M.J. Waldock, 2002. "Estrogenic Activity Measured in a Sewage Treatment Works Treating Industrial Inputs Containing High Concentrations of Alkylphenolic Compounds – A Case Study", Environmental Toxicology and Chemistry, ***21***(3): 507–514.
Snyder, S.A., D.L. Villeneuve, E.M. Snyder, and J.P. Giesy, 2001. "Identification and Quantification of Estrogen Receptor Agonists in Wastewater Effluents", ***Environmental Science and Technology***, ***35***(18): 3620–3625.
Sumpter, J.P., 2005. "Endocrine Disrupters in the Aquatic Environment: An Overview", ***Acta Hydrochimica et Hydrobiologica***, ***33***(1): 9–16.
Svenson, A., A.S. Allard, and M. Ek, 2003. "Removal of Estrogenicity in Swedish Municipal Sewage Treatment Plants", ***Water Research***, ***37***(17): 4433–4443.
Tilton, F., W.H. Benson, and D. Schlenk, 2002. "Evaluation of Estrogenic Activity from a Municipal Wastewater Treatment Plant with Predominantly Domestic Input", ***Aquatic Toxicology***, ***61***(3–4): 211–224.
Tyler, C.R., S. Jobling, and J.P. Sumpter, 1998. "Endocrine Disruption in Wildlife: A Critical Review of the Evidence", ***Critical Reviews in Toxicology***, ***28***(4): 319–361.

517b Superior Sorbents for the Removal of Pharmaceutical and Personal Care Products

Arturo J. Hernandez-Maldonado, Chemical Engineering, University of Puerto Rico at Mayaguez, PO BOX 9046, Mayaguez, PR 00681 and **Sindia Rivera-Jimenez**, Chemical Engineering, University of Puerto Rico-Mayagüez, PO BOX 9046, Mayaguez, PR 00681.

MCM-41 type mesoporous silicates were functionalized with nickel species via thermal monolayer and grafting techniques, respectively. The material potential to serve as sorbents for the selective removal of a pharmaceutical drug (i = Naproxen) from water was tested via single component adsorption equilibria measurements at ambient conditions and Ci = 18 ppm. In general, the grafting technique provided the best strategy to avoid metal leaching. Although the anchoring of the nickel species was successfully accomplished, analysis of the textural properties of the resulting sorbent indicated that a considerable reduction in average pore size (ca. 1 nm), and hence surface area, limited the ultimate uptake capacity. Nevertheless, after normalization with respect to accessible surface area, the grafted sorbents offered capacities comparable to those exhibited by a traditional activated carbon (0.42 vs. 0.67 μ mol/m^2). *Ab initio* and density functional theory calculations were employed to estimate the surface interaction energies for single sorbent clusters and attempt to describe the adsorption mechanism. Full optimization of the structures showed that the sorbate–sorbent interactions were at the weak-chemisorption level (~15 kcal/mol), which is suitable for ultrapurification applications and sorbent regeneration via simple engineering means. In addition, a natural bond orbital analysis indicated that the interaction between the Naproxen and the nickel functionality results in a redistribution of electrons within the most prominent orbitals in the transition metal center (i.e., 4s and 3d). This suggests that the interaction between sorbate and sorbent could generate a metal complex, but further studies are required to validate this hypothesis and to determine a sorbent regeneration strategy.

AUTHOR INDEX

AUTHOR INDEX

American Institute of Chemical Engineers
3 Park Avenue
New York, New York 10016-5991

CURRAN ASSOCIATES INC
proceedings.com